Spinach _{on the} Ceiling

The Multifaceted Life of a
Theoretical Chemist

Spinach on the Ceiling

The Multifaceted Life of a
Theoretical Chemist

Martin Karplus

Harvard University, USA & Université de Strasbourg, France

NEW JERSEY · LONDON · SINGAPORE · BEIJING · SHANGHAI · HONG KONG · TAIPEI · CHENNAI · TOKYO

Published by

World Scientific Publishing Europe Ltd.

57 Shelton Street, Covent Garden, London WC2H 9HE

Head office: 5 Toh Tuck Link, Singapore 596224

USA office: 27 Warren Street, Suite 401-402, Hackensack, NJ 07601

Library of Congress Cataloging-in-Publication Data

Names: Karplus, Martin, 1930–   author.

Title: Spinach on the ceiling : the multifaceted life of a theoretical chemist / Martin Karplus.

Description: New Jersey : World Scientific, [2020] | Includes bibliographical references and index.

Identifiers: LCCN 2019039400 | ISBN 9781786348029 (hardcover) |
    ISBN 9781786348067 (paperback) | ISBN 9781786348036 (ebook) |
    ISBN 9781786348043 (ebook other)

Subjects: LCSH: Karplus, Martin, 1930- | Chemists--Biography. | Nobel Prize winners--Biography.

Classification: LCC QD22.K315 K37 2020 | DDC 540.92 [B]--dc23

LC record available at https://lccn.loc.gov/2019039400

British Library Cataloguing-in-Publication Data

A catalogue record for this book is available from the British Library.

For any available supplementary material, please visit
https://www.worldscientific.com/worldscibooks/10.1142/Q0238#t=suppl

Typeset by Diacritech Technologies Pvt. Ltd.
Chennai - 600106, India

To my wife Marci and the Karplusians without whom this autobiography would not have been written.

Contents

———— ⊛ ————

*For instructions to download the figures in color and the films in Appendix 3, go to Appendix 5, Supplementary Material.

Preface

———— ⊕⊙⊛ ————

At 5:30 am on October 9, 2013, my wife, Marci was awakened by the telephone next to her side of the bed. Still half asleep, she woke me to say that the phone call was for me. My initial reaction was that if someone was telephoning at 5:30 in the morning, it was an emergency involving one of my children, and I expected the worst.[1]

When I answered the phone, the person at the other end of the line said, "This is Stockholm calling. Is this Professor Karplus?" I said yes, and someone (I later learned it was Gunnar Karlström, chairman of the Nobel Committee for Chemistry) congratulated me on winning the 2013 Nobel Prize in Chemistry. The other members of the committee then congratulated me in turn. It was only when Astrid Gräslund, the secretary of the committee, whom I knew from her work in biophysics, repeated the congratulation that I finally believed I had actually won the 2013 Nobel Prize, the highest recompense for scientific achievement.

At a celebration held in the Chemistry Department at Harvard later that morning, E.J. Corey who received the Nobel Prize in Chemistry in 1990 said how lucky I was that I only won the Nobel Prize when I was 83 years old, because I had had twenty more years than he had to work without being constantly interrupted by people wanting me to do this or that. His prediction was only too true.

I was born in Vienna, Austria, on March 15, 1930, and came to the United States as a refugee in October 1938. It seemed that I was destined to become a physician. For several generations, there had been one or more physicians in the family, partly because medicine was a profession in which Jews in Austria could work with relatively little hindrance from discrimination. Neither my brother nor any of my numerous cousins displayed any interest in becoming a doctor. I, by contrast, at the age of five went around bandaging chairlegs and other substitutes for broken bones. So my family concluded that I was to become "the" doctor of my

———— ⊛ ————

[1] My immediate reaction was that it must be something that had to do with my daughter Reba, who lived in Jerusalem. She had called on previous occasions at unusual times (given the seven-hour time difference) to assure me that she was fine after breaking news of a bombing. One time I remember distinctly: it was on September 4, 2001, when a bomb was detonated in a cafe in the center of Jerusalem near the Bikur Holim Hospital where she worked. I could hear the shots and screams over the phone with her. Nearly twenty individuals were hurt and Reba helped to treat many of them in the intensive care unit in Bikur Holim.

generation. My interest in medicine was reinforced by the stories about their work told by various relatives. Most fascinating were those of my Uncle Paul Wermer, a superb clinician, who would describe in detail how he went about diagnosing his patients. He made it sound like a detective story.[2]

I have sometimes wondered what my life would have been like if I had stayed in Vienna. I might well have done research, but I probably would not have had the same drive to accomplish something special as I did as a foreigner in the United States. Being a refugee and not quite belonging played a pivotal role in my view of the world and approach to science. It contributed to my realization that it was safe to stop working in fields that I felt I understood and to focus on different areas of research by asking questions that would teach me and others something new.

As I am reading over what I have written in the hope of finishing the book, I realize how fortunate I have been to arrive where I am now. This book describes my journey that began in Austria from where I escaped shortly after Hitler's entrance into Vienna, through my education in the United States, my postdoctoral time in England, and my faculty positions in Illinois, Columbia, Strasbourg, and Harvard. Although this brief summary of my career may make it appear that progress from one stage to the next was always easy, there were difficulties along the way. It was primarily by my optimistic outlook and faith in myself that I was able to overcome them, although sometimes it was a matter of luck. It is my hope that the description in this volume of the events that happened to me will aid young readers, in particular, to have a successful trajectory in their own lives.

What I have written provides at best only a fragmentary picture of my life, even of my scientific life. Missing are innumerable interactions, most of which have been constructive, although some not so, that have played significant roles in my career. The more than 250 graduate students, postdoctoral fellows, and visiting faculty who at one time or another have been members of the group now referred to as "The Karplusians" are listed in Appendix 1. Many have gone on to faculty positions and become leaders in their fields of research. They in turn are training students, so I now have scientific children, grandchildren, and great-grandchildren all over the world. I treasure my contribution to their professional and personal careers, as much as the scientific advances we have made together.

---

[2] In fact, he had a syndrome named after him, the "Wermer syndrome," which refers to rare diseases of endocrine glands.

The education of students in their formative years is a cardinal aspect of university life. My philosophy in graduate and postgraduate education has been to provide an environment where young scientists, once they have proved their ability, can develop their own ideas, as refined in discussions with me and aided by other members of the group. This fostered independence has been, I believe, an important element in the fact that so many of my students are now outstanding researchers and faculty members in their own right. My role has been to guide them when problems arose and to instill in them the necessity of doing things in the best possible way, not to say that I succeeded with all of them.

Along with my "scientific" family, my personal family has played an important role in my life. Reba and Tammy, my two daughters whose mother Susan died in 1982, both became physicians (thereby fulfilling my destined role). Reba lives in Jerusalem and Tammy lives on the west coast. My wife, Marci, and our son, Mischa, who has a degree from New York University School of Public Policy and a law degree from Boston University, complete my immediate family. As many people know, Marci also plays the pivotal role as my laboratory administrator, adding a spirit of continuity for the group and making possible our commuting between the Harvard and Strasbourg labs. Without my family, my life would have been an empty one, even with scientific success.

Most of the autobiography is based on work done while I was Professor of Chemistry at Harvard University and Professeur Conventionné at the Université de Strasbourg.

Acknowledgments

————⟡————

I would like to thank my editors at World Scientific Publishers for their aid, on the one hand, and patience, on the other, while I have been writing. Each of them, Koe Shi Ying, Ramya Gangadharan, and Britta Ramaraj, played an important role at various stages.

The encouragement I received from the publisher, Laurent Chaminade and Jennifer Brough, was essential in convincing me that expanding the Spinach on the Ceiling article into an autobiography was a worthwhile endeavor. In rereading my correspondence with Laurent, I was surprised to find that I only began working on the autobiography in August 2017, although it seems to me I have been working on the manuscript forever.

A very special thanks goes to Jane Sayers for reading the entire text and making numerous constructive suggestions on each of the chapters. Her aid in preparing the index was essential when I felt overwhelmed by the task.

Simone Conti was invaluable in helping me with problems that arose in using the Mac. He either knew the solution or, if not, had the confidence that enabled him to find it.

Chapter 1

My Ancestors

———— ⚛ ————

A t the time of my birth in 1930, all four of my grandparents and most of their extended families were living in Vienna. However, their origins and, in particular, the history of their arrival in Vienna were different. It is clear to me now that the intellectual traditions of my family, those of my grandfathers in particular, and also those of my aunts and uncles, as well as my parents, played an important role in the future I sought for myself.

My Mother's Family

My mother's father, Samuel (Samojla) Goldstern, was born on December 31, 1865, and grew up in Odessa (Figures 1.1 and 1.2).

In the late 1800s Odessa was the fourth largest city in Russia and a thriving market town, sometimes referred to as the Russian Marseille. My grandfather's parents were wealthy grain merchants and part of the large Jewish community, which made up about 30% of Odessa's population. The family had moved to Odessa from Lemberg (Lviv) in Galicia, a province of the former Poland, now part of the Ukraine. Under the reign of Joseph II (1780–1790), Jews were permitted to live in towns that were part of the Hapsburg Monarchy. Prior to this, Jews could work in the towns but had to live outside.

Figure 1.1. My maternal grandparents, Marie (Bernstein) Goldstern and Samuel Goldstern

Figure 1.2. My grandmother with her four children: (left to right) my mother Lucie, Claire, Alex, and Lene

The name Goldstern first appeared when Joseph II gave the order that Jews should have surnames, in part to make taxation easier. Previously Jews, like many other ethnic groups, usually named the eldest son after the father: if the father was Mendel, the son was Mendelson. Names were sold to Jews, with names such as diamond, gold, and silver being expensive. Clearly Goldstern was one of these. Poor Jews were often given unpleasant names such as Schmerz, which means pain in German.

The first head of the family with the name Goldstern was Rabbi Mendel Goldstern, several previous generations having named the eldest son Mendelson [**Norbert Goldstern, private communication**].[1]

Abraham, the youngest son of Rabbi Mendel Goldstern and the father of my grandfather, was born in 1832 in Lemberg and died in 1905 in Vienna. Abraham was the only Jewish wholesaler in Odessa allowed to deal in the "first" class, the one class permitted to operate at commodity markets and exchanges as well as to export. He owned around forty buildings in the center of Odessa, including the main post office, gymnasium, and lyceum. Having come from Lemberg, the family still spoke German in Odessa, rather than Yiddish or Polish. The family

---

[1] Rabbi Mendel, the great-grandfather of Samuel Goldstern, lived near the end of the 18th century and was a Chassidic rabbi; his son, also named Mendel, was a Talmudic scholar and banker or, more specifically, a moneylender.

were what was colloquially called "Christmas Tree Jews"; i.e., "assimilated Jews," who no longer practiced the Jewish religion, a far distance from the rabbinical origins of the family.

When the first marriage of Abraham Goldstern was without offspring after ten years, it was dissolved in about 1860 and he married Marie Kitower, a sixteen-year-old girl. That he was over thirty years old may appear shocking to us now, but it was not unusual at the time. Marie gave birth to a child every year, and had a total of sixteen, of whom three died in early childhood. This was an exceptionally low death rate for the time, due probably to the parents' very good financial position and the care provided for their children. They were chauffeured daily to Jewish schools and both the boys and girls received a secondary education. Of the thirteen who survived infancy, eight were boys and five were girls.

Figure 1.3. A painting, which I owned, by the Israeli artist Tumarkin of a famous scene from the Eisenstein film Potemkin

In October 1905, the economic downturn at the time and rising anti-Semitism resulted in a bloody pogrom under Czar Nicholas II. This was caused in part by the revolt of the sailors on the battleship Potemkin, immortalized by the Sergei Eisenstein film (Figure 1.3).

The evening before the slaughter of Jews began, my grandfather and some of his siblings were still in Odessa, but they managed to escape to Vienna. These events presaged what happened to me and my parents when Hitler entered Austria in 1938.

Samuel (Samojla) Goldstern, the third oldest sibling, began his medical studies in Odessa, as did his older brother Sima. Samojla completed his doctorate at the University of Vienna Medical School in 1892, where he specialized in internal medicine. In Vienna he met his future wife Marie (Mania) Bernstein. She was born in 1877 in Vinnytsia, a city located in central Ukraine, about 200 miles inland from Odessa. Manja also had moved to Vienna where she studied economics at the University of Vienna. They were married in 1899, only after she had completed her studies as wished by her parents.

Samojla's first position was as a physician at the Fango Heilanstalt (Figures 1.4, 1.5a, and 1.5b), a private clinic that was founded in 1892 and briefly located at Brünlgasse 12. In 1898, it moved to its permanent location at Lazarettgasse 20, where it occupied an entire building. Samojla purchased the Fango Heilanstalt in 1915 and continued to run it until shortly before he died in March 1939.[2] I often visited the Fango because my grandparents had their apartment on an upper floor. The private clinic specialized in the treatment of rheumatological diseases by use of a mildly radioactive mud.[3] I had always assumed that "Fango" was a made-up name and learned only decades later that *fango* is nothing more than the Italian

Figure 1.4. Present-day photo of the Fango Heilanstalt

---

[2]Their son Alex, my uncle, was interned in the Dachau concentration camp in the spring of 1938. My grandfather was told by a Nazi official that Alex would be released if he signed the Fango over to the Nazis. Of course, my grandfather had no choice and signed, legalizing the Nazi takeover. Alex was released as a result in January 1939, with the condition that he leave Austria within three weeks. It was impossible to obtain a visa to the United States in such a short time, but it was possible to obtain one for New Zealand. Consequently, Alex moved to New Zealand and lived there until he returned to Vienna after the war. He reclaimed the Fango and ran it until retiring, by which time the Fango had become part of the Viennese hospital system, which it still is today.

[3]Interestingly, in October 2016, an article in *Arthritic Care and Research* reported a randomized trial showing that "mud bath therapy" reduces pain and improves function in patients with osteoarthritis of the knee. Perhaps the small radioactivity was not essential.

word for *mud*. Many of the patients at the Fango Heilanstalt were wealthy Arabs from the Middle East, an example of how the history of Arabs and Jews has long been intertwined and not always in the present destructive manner.

Figure 1.5a. Description of the Fango, from the time when my grandfather Goldstern was the chief physician

Samuel's brothers and sisters were also highly educated and had successful lives. The eldest, Mischa, was an accountant and is the source of our son's name.[4] Sima, the second oldest, like my grandfather, began his medical studies in Odessa and completed them in Vienna. He then switched to chemistry and continued to live in Vienna even after the Nazis took over. Learning that he was going to be deported to a concentration camp in 1942, he committed suicide. Two brothers, Philip and David, also studied chemistry; they moved to Bucharest, Romania, where they worked together and founded a successful petroleum company.

Among the five sisters, the youngest, Eugenie, nicknamed Jenja, was an ethnologist [**Ottenbacher, 1999**] (Figure 1.6).

Rather than focusing on primitive civilizations in distant lands as did most ethnologists, she studied peoples in isolated alpine communities in Europe. She donated her collection of children's toys to the Ethnographic Museum of Vienna. It was largely ignored until French ethnographers became interested in her study of Bessans, a French village, which was still unspoiled by tourism when she spent two winters there just before the First World War.

[4]Marci and I had a difficult time choosing a name for our son. Three weeks before the predicted due date, Marci visited the Paris clinic, where Mischa was born, to have an ultrasound examination. The exam verified that we were going to have a boy. As we were going down in the elevator after the exam, we immediately agreed on the name Mischa, in part because the name would work well in French and English.

1904 *1929*

*Zur Erinnerung
an das 25 jährige Jubiläum
meiner Anstaltsleitung*

Figure 1.5b. 25th Anniversary of the Fango

Figure 1.6. Eugenie Goldstern

I had been completely unaware of Aunt Jenja's work until one day when Marci and I visited a bookstore in Annecy, the nearest large town to Manigod, where we have spent our summers since 1970. Marci noticed a small book by a Eugenie Goldstern and in reading her description, we realized that she was my great aunt. The book was a French translation by Francis Tracq of her PhD thesis (Figure 1.7).

When we visited the village of Bessans in the early 1980s we saw her book on the counter of the local newspaper store. Marci proudly told the owner that Eugenie was my great aunt. Within a few minutes, even before we reached the bed-and-breakfast where we were staying, the whole village was aware of our presence. Apparently we were the first members of the family to visit Bessans since Eugenie had fled over a mountain pass to Italy at the beginning of the First World War.

EUGENIE GOLDSTERN

BESSANS

La vie dans un village de Maurienne

PRESENTATION DE FRANCIS TRACQ

les Savoisiennes
CURANDERA

Figure 1.7. Cover of the French translation of Eugenie Goldstern's PhD thesis

My memories of Aunt Jenja are from the time in Vienna when I would visit my grandparents' apartment in the Fango on weekends for typical Viennese afternoon teas with wonderful cakes. Aunt Jenja, who also lived in the Fango, usually brought some toys for my brother and me to play with. She remained in Vienna after the Nazis took over and was deported to the concentration camp in Sobibor where she was murdered.

Sonia, the second youngest daughter, married Leopold Wermer, a dentist. Their son was Paul Wermer, a physician whose stories about his work inspired me, as I already mentioned in the Preface. Hans (Johnny) Wermer, the son of Paul and his wife Eva, was my brother's age and we often played together in Vienna. Johnny and his parents were able to immigrate to the United States in 1939.

My Father's Family

My paternal grandfather's family had its origins in the village of Hotzenplotz, which is located about 150 miles from Prague. It was then part of the Austro-Hungarian Empire and is now in the Czech Republic. Hotzenplotz was almost entirely Roman Catholic, but it did have a small Jewish population and an old Jewish cemetery still exists there.[5]

The Karplus family continued living in Hotzenplotz and was primarily involved in commerce, particularly based on lumber for use in buildings and railway tracks. Gottlieb Karplus, my great grandfather, was born in Hotzenplotz, but he and his wife Elizabeth moved to Vienna in 1859. Elizabeth died in 1925 in the Fango Heilanstalt, the Goldstern clinic. My grandfather Johann Paul Karplus (Figures 1.9 and 1.10) was professor of neurology at the University of Vienna where he had received his medical doctorate. He is well known for his discovery of how the hypothalamus functions. In experiments he conducted with his coworker Kreidl in the early 1900s, he showed that an electrical stimulation of the hypothalamus

[5]My wife Marci and I were in Prague as part of a lecture trip in 1988. While visiting a museum next to the cemetery (Figure 1.8a), we saw an exhibition of children's drawings from the Theresienstadt concentration camp. More than 15,000 children lived there for months before being transported to death camps. An art teacher, Friedl Dicker-Brandeis, organized drawing classes for the children to make their lives more bearable. When we looked at the drawings (Figure 1.8b), we were shocked to find some by a girl named Erika Hanna Karplosová, the Czech version of Karplus. She was born in Brno, which had a large Jewish population before the war. It is a city located midway between Hotzenplotz and Vienna. Erika is a distant cousin, whose birthday was April 1, 1930. She would have been the same age as I am had she not died in Auschwitz in 1944.

Figure 1.8a. Jewish cemetery with museum building in Prague

released a fluid, which transmitted the information of "fright" to the rest of the body. It is now known that, although the hypothalamus is part of the brain, it has endocrine glands that release neurohormones.

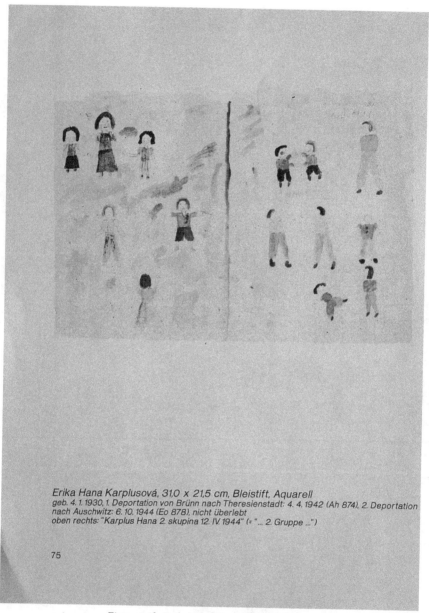

Erika Hana Karplusová, 31,0 x 21,5 cm, Bleistift, Aquarell
geb. 4. 1. 1930, 1. Deportation von Brünn nach Theresienstadt: 4. 4. 1942 (Ah 874), 2. Deportation
nach Auschwitz: 6. 10. 1944 (Eo 878), nicht überlebt
oben rechts: "Karplus Hana 2. skupina 12. IV. 1944" (= "... 2. Gruppe ...")

75

Figure 1.8b. Erika's drawing in the exhibition

My grandfather Johann Paul married Valerie von Lieben, who belonged to an even wealthier Viennese Jewish family. They lived in one of the "Ringstrasse Palaces." These were imposing buildings, many of which had been constructed by wealthy Jewish families, who had been granted the right to own property in

Figure 1.9. My paternal grandparents: Johann Paul Karplus and Valerie (von Lieben) Karplus

Figure 1.10. My paternal grandparents with their four children, 1930: (left to right) Edu, Johann Paul, Walter, Valerie, Heinrich, and my father Hans

Vienna in the 1850s when Emperor Franz Joseph I (1830–1916) initiated the Ringstrasse project. He had the wall surrounding Vienna torn down with the aim of making the Ringstrasse the finest avenue of Vienna. His plan also included using the money raised from the sale of the land for the construction of the State Opera, the State Art Museum (Kunsthistorisches Museum), and the State Theatre. Perhaps the most well known of the Ringstrasse Palaces is Palace Ephrussi, made

Figure 1.11. Present-day photo of Auspitz-Lieben Palace with Cafe Landtmann on street-level

famous by the book *The Hare with Amber Eyes*. The Auspitz-Lieben Palace (Figure 1.11) of my family was built in 1872; the Cafe Landtmann, which is still one of the most elegant Vienna coffeehouses, was opened on the ground floor in 1873. I remember visiting my grandparents in their apartment in the Auspitz-Lieben Palace on the first floor (the second floor for American readers), which was crammed full with furniture and works of art.

Figure 1.12. Self-portrait by Marie-Louis Motesiczky

On my grandmother, Lieben's, side of the family there were several well-known scientists. The best known is Robert von Lieben, a brother of my grandmother. (The "von" was an honorific title of nobility bestowed on wealthy families, including wealthy Jewish families, for contributions to the Austro-Hungarian Empire). He invented the radio diode, which he patented in Europe. There were also several chemists in the Lieben family, with the men in each generation divided between those who were scientists and those who went into business. Henriette, a sister of my grandmother Valerie, married Edmund von Motesiczky, a member of another wealthy Jewish family. Henriette was an amateur painter. The Motesiczkys had two children: Karl, a gifted musician, who died in Auschwitz when he was 39 years old; and

Marie-Louis, who was a student and a muse of Max Beckman, and became a well-known painter in her own right, although she was only recognized late in life (Figure 1.12). Marie-Louis escaped with her mother to England during the war.

The Lieben family was honored by an exhibition in 2004–2005 at the Viennese Jewish Museum entitled "The Liebens. 150 years of a Viennese Family" (Figure 1.13).

As was described there, members of the family made many contributions to life in Vienna. Among others is the foundation of the Ignaz Lieben Prize in 1862, which was awarded for an outstanding contribution to science in Austria. It was then referred to as the "Austrian Nobel Prize."[6] In 1937, shortly before the Nazis took over Austria, the last original Ignaz-Lieben Prize was awarded. After the war, the Ignaz-Leiben Prize was reinstituted in 2004, thanks to a donation from Dr. Alfred Bader who cofounded the Aldrich Chemical Company, and his wife Isabel.

Both my parents had been physics students at the University of Vienna, though neither had finished their degree. During the 1930s my father Hans worked in the banking business that had its origins on the Lieben side of the family, and my mother Isabella (Lucie) was the dietician at the Fango Heilanstalt, her father's clinic.

---

[6]In fact, several Nobel Prize winners, including Otto Loewi and Karl von Frisch, had received the Lieben Prize, as did Lisa Meitner, a distant cousin who did not receive a Nobel Prize; in the opinion of many scientists she should have received one jointly with Otto Hahn (Nobel Prize in Chemistry, 1944) for her theoretical contribution to understanding nuclear fission.

Figure 1.13. Cover of the Catalogue of "The Liebens" exhibition

Chapter 2

Childhood Years in Europe

y childhood home, where we lived until 1938, was located in the Viennese suburb named Grinzing, which was reputed as a wine-growing area. There were small informal inns (Heurige), where we sometimes went in family groups for relaxed evenings of eating, and also drinking (mainly the adults) the fruity young white wine of the region (Figure 2.1).

The Heurige were also great for us children because in between eating sausages, cheese, and bread, we could play in the garden.

We had a modest house set in a garden with a small pool where my brother and I (Figure 2.2), as well as neighborhood friends, could splash during hot summer months (Figure 2.3).

Figure 2.1. A recent picture of a typical Heurige

Figure 2.2. My brother and I as children

Figure 2.3. My childhood home today

In 1935, when I was five and my brother Robert was eight, several memorable events took place. The garden to the house was fronted by a high metal fence with spikes on top (Figure 2.4). Robert, who was rather wild, liked to show off by climbing on the fence. One day he slipped and fell so that one of the spikes penetrated his chin. I remember being terribly frightened by the scene. His screams were heard by Mitzi, who was our beloved nanny. Mitzi was able to lift Robert off the fence. He had been extremely lucky in that, miraculously, very little damage had been done. Moreover, it did not stop my brother from continuing his habit of showing off before his friends. Later that same year, we were playing in a neighborhood park where we liked to climb the trees. That day we were there with cousin Johnny Wermer, and Robert climbed high up on a tree. It was a very windy day and Robert fell out of the tree, hitting his head when he landed on the ground. Although he was conscious, he was unable to get up. Johnny ran back to our house, which was only a few minutes away and told my parents. They in turn called Uncle Alex, who had a motorcycle with a sidecar (Figure 2.5), and he took Robert to the hospital where he remained for about a week.

The whole family was extremely concerned, and our home, usually lively, was very quiet all week long. When Robert returned from the hospital, he did not remember what had happened. The only obvious consequence was that he was deaf in his right ear because his eardrum had been ruptured by his fall. What gradually became evident was the change in his personality. Instead of being a

Figure 2.4. Our house with my brother and me in 1935

wild show off, Robert became quieter and more studious. Thus, one might say that the fall had a "positive effect" on his future. He became an outstanding student, both in Austria before we left and later when we were in the United States.

I am told that one of my initial reactions when Uncle Alex returned from taking Robert to the hospital was to say that now I did not have to compete with him to ride in the sidecar of the motorcycle. This presaged my relationship with my

Figure 2.5. Alex's motorcycle with my brother and me plus our cousins Gus and George

brother throughout my youth. I was always trying to keep up with him and his friends, never seeming to realize that being three years younger often made it impossible.

In the early 1930s, when owning an automobile was still relatively rare, we already had a small car, a "Steyr Baby." One day when it was parked in front of our house, I scooted into the driver's seat and pretended to drive. I inadvertently released the brake, and the car started to roll downhill. I was terribly frightened as the car approached a pit at the end of the street. Somehow, I managed to steer the car, so it turned just before reaching the pit and stopped. I recall the slope of the hill as being steep and the pit being very deep. On a visit to Vienna in 1981, we went to see my childhood home. I discovered that the "steep" hill was a relatively gentle slope and the "pit" a relatively shallow ditch.

Our house had been confiscated by the Nazi regime during the Second World War and "sold" to a non-Jewish family. When we visited Vienna in 1981, it was occupied by the family of the son of the couple who had acquired it from the Nazis. He invited us into the house, but was clearly embarrassed by our presence. Nevertheless, he showed us around and pointed out where a bomb had damaged the living room. When Marci asked about more details of the bombing, the owner became defensive, saying "I was too young to remember." My parents never tried to get the house returned to them. Nor did they try to reclaim the "Steyr Baby," which had likewise been confiscated. The only restitution we received was 5080 Euros, more than 25 years after my parents died, when I learned that it was available to any Austrian who left because of the possibility of persecution.

Stories from my early childhood, most of which I know from their retelling by my parents, aunts, and uncles, indicate that I was a strong-willed independent child (in a positive sense) and a brat (in a negative sense). One such story concerns my "escape" at the age of three from a summer day care, where I apparently did not like the way I was being treated. One morning I simply walked out and somehow, despite the center being more than a mile from my house, made my way home. Both the day care staff and my parents, who had been notified, were extremely worried about my disappearance and were searching for me. My parents were so happy to see me that my punishment only consisted of a mild scolding. The venture was justified in my mind because after that I was allowed to stay home.

Then there was the infamous "spinach incident." Mitzi told me that I must eat my spinach. (Popeye did not exist in Austria, but unfortunately spinach did.) With all the vehemence I could muster, I took a spoonful of the spinach and threw it at the ceiling. The spinach stain remained visible on the ceiling for a long time and was pointed out at appropriate moments when my parents wanted to indicate what a naughty child I was.

In a description of me, written by my mother in 1936 as part of a course she was taking, she comments "Before going off to Kindergarten, Martin walks around and talks. He really talks a lot, he comes to me in the morning to tell me what a capable child he is, what he has already accomplished, and what his plans are for the day.... It is very important to him that his mother is listening, and he notices immediately when she is not. He says 'Mama you are not listening' and may repeat what he said three times."

Another experience that has stayed with me from kindergarten has to do with my lack of musical ability. One day as the class sang together, the teacher came up to me and softly said that it would be better if I only pretended to sing since my enthusiastic participation was invariably off-key. To this day, I follow this "rule." None of my children seem to have escaped this "inheritance," though my granddaughter Rachel apparently has done so; she sings as part of a school choir that won a state competition.

During my early childhood years, we traditionally took hikes in the surroundings of Vienna and summered at an Austrian lake or on the Adriatic coast of Italy with several families that were either relatives or friends with children of similar ages (Figure 2.6).

Such extended family activities were an integral part of my growing up and gave me confidence and a sense of belonging (Figures 2.7 through 2.9).

Figure 2.6. The extended family with Robert and me in center

Figure 2.7a. My father on a bicycle with Robert and me, my cousin Johnny, and his nanny Besse

Figure 2.7b. My mother, Robert, Johnny, and I

Figure 2.8. Robert and I in a kayak

Figure 2.9. Steering a sailboat on Attersee

One day at the beach, a friend of my parents picked me up and cuddled me, much to my dismay. I yelled out, "*Ich bin ein Nazi*" ("I am a Nazi"), which so shocked her that she dropped me. Clearly, I had somehow realized, presumably from listening to my parents and others, that being a Nazi was the worst possible thing to be (Figure 2.10).

Figure 2.10. In the arms of a relative with my brother Robert and my cousin Gus standing in the foreground

Chapter 3

Final Days in Europe

———————— ⚬✛⚬ ————————

O ur life was already changing significantly before the Nazis entered Austria in 1938, even from the viewpoint of an eight-year-old. Among our neighbors were two boys of comparable ages to my brother Robert and me. They were our "best friends," and we played regularly with them, including splashing in the small pool in our garden. In the spring of 1937, they suddenly refused to have anything to do with us and began taunting us by calling us "dirty Jew boys" when we foolishly continued to try to interact with them. Similar problems occurred at school with our non-Jewish classmates. Before this, my school experience in the first grade (Figures 3.1a and 3.1b) and the beginning of second grade had been wonderful, in part because I had a great teacher, Herr Schraik, not the least of whose outstanding attributes was that his wife ran a candy store. When my class was ready to advance to second grade, the parents petitioned that Herr Schraik be "promoted" with us and, because of his outstanding record as a teacher, this request was granted. Nevertheless, in the middle of that school year (1937–1938), he was no longer allowed to teach. He was Jewish and the authorities had decided that any contact with him would contaminate the minds of the children. The new teacher was incompetent and blatantly anti-Semitic—he constantly criticized the Jewish students like myself, independent of how well or poorly we were doing. The situation soon became so bad that my parents took me out of the school.

There were a number of differences between elementary school in Austria and in America. In Austria, school was Monday through Saturday, with homework starting in the first grade, and no extracurricular activities. My class had only boys; girls attended separate classes at the same school (Figure 3.2).

Perhaps the most significant difference of my schooling in Austria had to do with my being left-handed. When I started first grade, I was obliged to learn to write with my right hand. Whatever the supposed psychological consequences of that may be, I have always been grateful that right-handedness in writing was imposed on me. This was particularly true when I first went to the United States and saw the contortions children went through to write with their left hand. Clearly everything, at least in Western European languages, is set up for right-handed people.

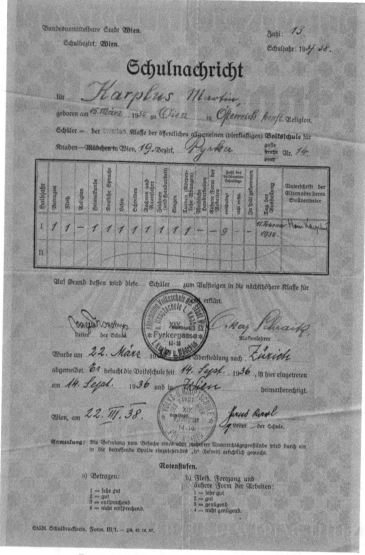

Figure 3.1a. My report-card in the first grade

On March 12, 1938, the German troops crossed the border into Austria and completed the *Anschluss,* the "joining" of Austria with Nazi Germany, which had been specifically forbidden by the Versailles Treaty after the First World War. The Germans had been "invited" into Austria by the puppet Seyss-Inquart, who took over the Austrian government after Chancellor Schuschnigg was forced to resign. The night before the Nazi troops entered the city, my family and some

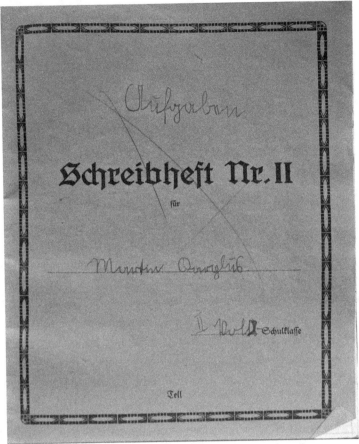

Figure 3.1b. Cover of my exercise book

friends were listening to the radio in the living room, which was darkened to conform to the curfew and black-out requirements in case of air attacks. I was furious, not because of what was happening politically, but rather because my parents had planned an early birthday party for me that evening (my actual birthday being March 15), and I was far from being the center of attention. Hitler entered Austria in triumph and his troops were welcomed by enthusiastic crowds. (To this day, more than 75 years later, I have mixed feelings about visiting Austria because anti-Semitism is still prevalent. However, when I learned in 2012 that I am still officially an Austrian citizen, I did not "resign." Having dual nationality is of importance because as a citizen of Austria, I am part of the European Union.)

Figure 3.2. Elementary school students with me in the foreground

My parents had been concerned about the rise of anti-Semitism and the possibility of Hitler's takeover of Austria for some time. Several unsettling events had already occurred in Austria. In the early 1930s, Hitler had expressed his dream of uniting all German-speaking people, primarily Germany and Austria, and in 1934, the head of the Austrian government, Chancellor Dollfuss, had been assassinated by Austrian Nazis.

Given their awareness of the political situation, my parents had begun to prepare for us to emigrate. For the previous three years, Aunt Claire, who had studied in England, had been teaching English to Robert and me. On February 21, 1938, I received my passport (Figure 3.3), and well before March 12, train tickets for a "ski trip" to Switzerland had been purchased and a bed-and-breakfast "pension" had been reserved in Zurich.

A few days after the *Anschluss*, my mother, brother, and I left Austria by train for Switzerland. That day when I came home I saw that all the trunks were out. I was told to put on my warm coat and get into the car. We then were driven to the train terminal and were hustled onto the train. Only then was I told what was happening: we were going to Switzerland to escape from Austria and the Nazis.

The most traumatic aspect of our departure was that my father was not allowed to come with us and had to give himself up to be incarcerated in the Viennese city jail. In part, he was kept as a hostage so that any money we had would

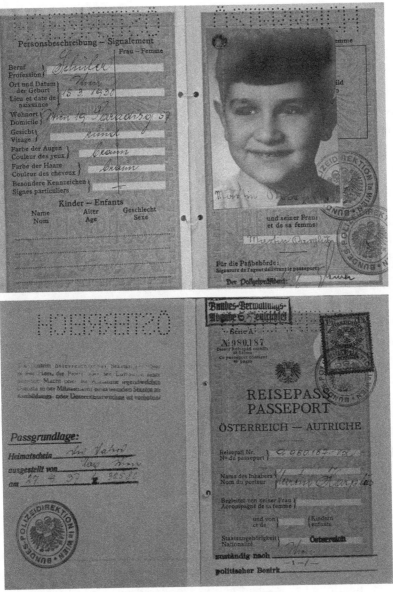

Figure 3.3. My passport from 1938

not be spirited out of the country. My mother reassured my brother and me, saying that nothing would happen to him, though of course she herself had no assurance that this was true. At that time, the Nazi government still allowed Jews to leave Austria, as long as they left their money and possessions behind. One way of surreptitiously getting money out of the country was to buy diamonds and hide them in clothing and food. During our train trip to Zurich, we were held up at the border and everyone had to get out of the train with their suitcases. After ours were examined and nothing was found, we were allowed to board the train again. My mother had packed a lunch, which included a large sausage. The guards meticulously reduced it to thin slices, presumably searching for hidden diamonds. Fortunately, the sausage could still be eaten, and Robert and I were not too bothered by this event.

My father, Hans, was the eldest of the four brothers and was the only one still in Vienna at the time of the *Anschluss*. The next oldest, Eduard (Edu), an engineer, had already emigrated to the United States some years before. The third brother, Heinrich, who had graduated from the medical faculty in Vienna in 1930, where he had studied to be a pathologist, had emigrated to Israel in 1936. The youngest brother, Walter, had arrived in the United States prior to 1938. After the death of my grandparents, my father was in the process of settling their estate when the Nazis took over Austria. While my mother, brother, and I were allowed to leave Austria within days of the *Anschluss*, my father was forced to stay behind. He was kept in jail until he had signed over all our belongings and assets to the Nazis. This involved, among other items, our house in Grinzing and our "Steyr Baby" car.

I learned many of these details only as the result of a serendipitous event. Joseph Ostermayer, the Austrian Federal Minister of Arts and Culture, had been invited to present the keynote lecture at the March 2014 Conference on the "Transformation of Intractable Conflicts" held at Harvard University's Herman C. Kelman Institute. During Dr. Ostermayer's stay in Boston, the Austrian Consul arranged for him to see me to congratulate me on my Nobel Prize. While we were chatting, he asked (I had the impression more out of politeness than anything) "Is there anything I can do for you?" I have never been one to let such an opportunity go by and mentioned that I had always wondered what had happened to my father when he was jailed in Vienna. Dr. Ostermayer said he would look into what had happened and, true to his word, I received a complete record of everything that had been confiscated, the Nazis having kept a detailed record, which was preserved in the Austrian State Archive. Apparently, my father was jailed by the Gestapo on April 3, 1938, and a detailed listing of our possessions had been made by them (Figure 3.4).

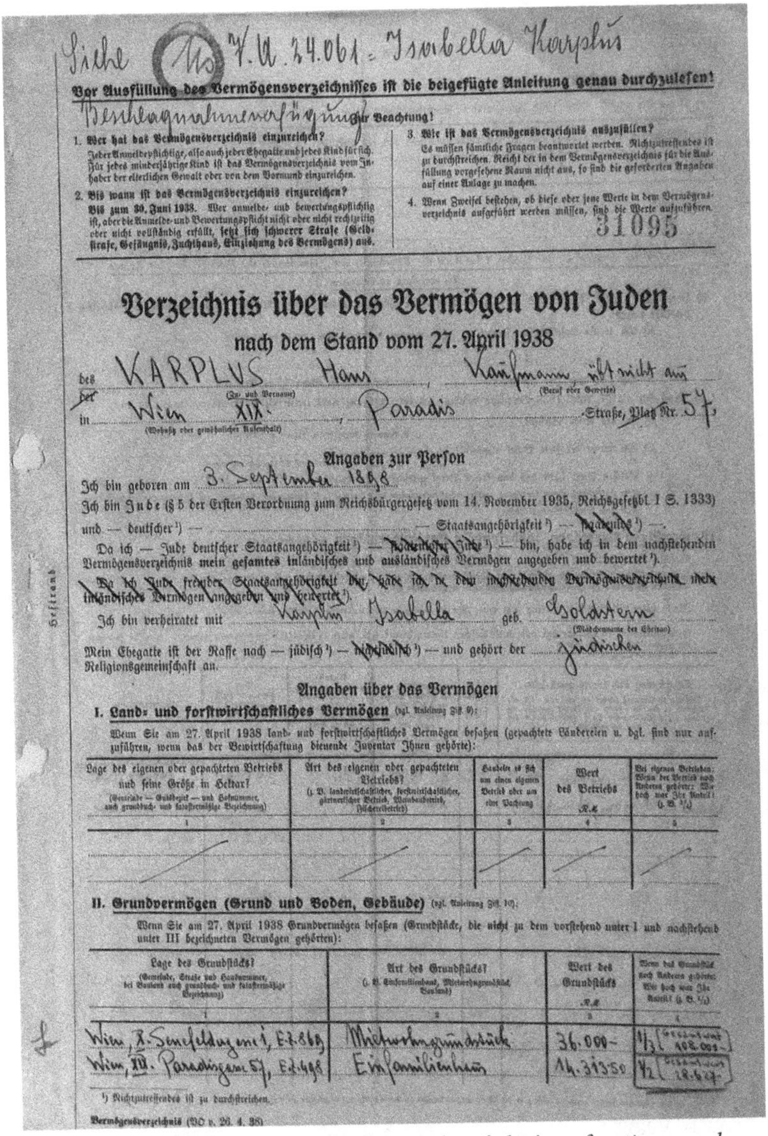

Siehe V.A. 24.061 = Isabella Karplus

Vor Ausfüllung des Vermögensverzeichnisses ist die beigefügte Anleitung genau durchzulesen!

Beschlagnahmeverfügung

Zur Beachtung!

1. **Wer hat das Vermögensverzeichnis einzureichen?** Jeder Anmeldepflichtige, also auch jeder Ehegatte und jedes Kind für sich. Für jedes minderjährige Kind ist das Vermögensverzeichnis vom Inhaber der elterlichen Gewalt oder von dem Vormund einzureichen.
2. **Bis wann ist das Vermögensverzeichnis einzureichen?** **Bis zum 30. Juni 1938.** Wer anmelde- und bewertungspflichtig ist, aber die Anmelde- und Bewertungspflicht nicht oder nicht rechtzeitig oder nicht vollständig erfüllt, **setzt sich schwerer Strafe (Geldstrafe, Gefängnis, Zuchthaus, Einziehung des Vermögens) aus.**
3. **Wie ist das Vermögensverzeichnis auszufüllen?** Es müssen sämtliche Fragen beantwortet werden. Nichtzutreffendes ist zu durchstreichen. Reicht der in dem Vermögensverzeichnis für die Ausfüllung vorgesehene Raum nicht aus, so sind die geforderten Angaben auf einer Anlage zu machen.
4. Wenn Zweifel bestehen, ob diese oder jene Werte in dem Vermögensverzeichnis aufgeführt werden müssen, sind die Werte aufzuführen.

31095

Verzeichnis über das Vermögen von Juden

nach dem Stand vom 27. April 1938

des KARPLUS Hans (Zu- und Vorname), Kaufmann (Beruf oder Gewerbe)

in Wien XIX. (Wohnsitz oder gewöhnlicher Aufenthalt), Paradis Straße, Platz Nr. 57

Angaben zur Person

Ich bin geboren am 3. September 1898

Ich bin Jude (§ 5 der Ersten Verordnung zum Reichsbürgergesetz vom 14. November 1935, Reichsgesetzbl. I S. 1333) und — deutscher[1]) — Staatsangehörigkeit[1]) — staatenlos[1]) —.

Da ich — Jude deutscher Staatsangehörigkeit[1]) — staatenloser Jude[1]) — bin, habe ich in dem nachstehenden Vermögensverzeichnis **mein gesamtes inländisches und ausländisches** Vermögen angegeben und bewertet[1]).

Da ich Jude fremder Staatsangehörigkeit bin, habe ich in dem nachstehenden Vermögensverzeichnis mein inländisches Vermögen angegeben und bewertet[1]).

Ich bin verheiratet mit Karplus Isabella geb. Goldstern (Mädchenname der Ehefrau)

Mein Ehegatte ist der Rasse nach — jüdisch[1]) — nichtjüdisch[1]) — und gehört der jüdischen Religionsgemeinschaft an.

Angaben über das Vermögen

I. Land- und forstwirtschaftliches Vermögen (vgl. Anleitung Ziff. 9):

Wenn Sie am 27. April 1938 land- und forstwirtschaftliches Vermögen besaßen (gepachtete Ländereien u. dgl. sind nur aufzuführen, wenn das der Bewirtschaftung dienende Inventar Ihnen gehörte):

Lage des eigenen oder gepachteten Betriebs und seine Größe in Hektar? (Gemeinde — Gutsbezirk — und Hofnummer, auch grundbuch- und katastermäßige Bezeichnung)	Art des eigenen oder gepachteten Betriebs? (z. B. landwirtschaftlicher, forstwirtschaftlicher, gärtnerischer Betrieb, Weinbaubetrieb, Fischereibetrieb)	Handelt es sich um einen eigenen Betrieb oder um eine Pachtung	Wert des Betriebs RM	Bei eigenen Betrieben: Wenn der Betrieb auch Anderen gehörte: Wie hoch war Ihr Anteil? (z. B. 1/3)
1	2	3	4	5
/	/	/	/	/

II. Grundvermögen (Grund und Boden, Gebäude) (vgl. Anleitung Ziff. 10):

Wenn Sie am 27. April 1938 Grundvermögen besaßen (Grundstücke, die nicht zu dem vorstehend unter I und nachstehend unter III bezeichneten Vermögen gehörten):

Lage des Grundstücks? (Gemeinde, Straße und Hausnummer, bei Bauland auch grundbuch- und katastermäßige Bezeichnung)	Art des Grundstücks? (z. B. Einfamilienhaus, Mietwohngrundstück, Bauland)	Wert des Grundstücks RM	Wenn das Grundstück auch Anderen gehörte: Wie hoch war Ihr Anteil? (z. B. 1/3)
1	2	3	4
Wien, X. Senefeldergasse 1, E.Z. 869	Mietwohngrundstück	36.000.–	1/3 (Gesamtwert 108.000.–)
Wien, XIX. Paradisgasse 57, E.Z. 498	Einfamilienhaus	14.313.50	1/2 (Gesamtwert 28.627.–)

[1]) Nichtzutreffendes ist zu durchstreichen.

Vermögensverzeichnis (VO v. 26. 4. 38)

Figure 3.4. First page of the Nazi listing of my father's confiscation record

Although much had been confiscated, my father had been able to ship a container to the United States containing furniture and other belongings from my paternal grandparents' apartment in the Auspitz-Lieben Palace (Figure 3.5).

He had shipped four containers, one for each of the four brothers, and they all reached their destinations. Along with heirloom dishware and linen, ours

Figure 3.5. A typical room in the Auspitz-Lieben Palace

contained a contemporary copy of the Veronese painting of "Moses in the Bulrushes" (Figure 3.6), as well as a few pieces of furniture that had been in the family for generations. Two of these were donated by Marci and me to the Jewish Museum in Vienna. One is an upholstered rocking-chair with carved swans for armrests; another is a folding card table (Figure 3.7) on which my grandfather played Taroc with Sigmund Freud, who was a colleague and friend of the family. Freud's first patient, described under the pseudonym, Frau Caecilie M., was Anna von Lieben, my great-grandmother. Anna had five children, was outstanding at chess, and like many highly intelligent women of that period, was supposed only to marry and raise a family. Women were rarely allowed to work and in some cases were described as "hysterical." Anna was one such woman and she was treated by Freud for many years.

My parents hoped to go to the United States, but visas to the United States were granted only to applicants who had an "affidavit," a document from an American citizen guaranteeing their financial support.[1] My Uncle Edu had become chief

---

[1]Many Jews were allowed to leave Austria (or Germany) if they had a visa to enter another country. Some ended up in South America, Australia, and New Zealand, but, as history has recorded, many were not able to reach a safe country. Of the 180,000 Jews in Austria before the war, more than 120,000 were able to emigrate, but many went to other European countries only to be caught by the Nazis there and died in concentration camps.

Figure 3.6. "Moses in the Bulrushes"

engineer at the General Radio Corporation in Boston, where he invented the Variac, which is still widely used for continuously varying the electric voltage. The president of General Radio, Mr. Eastman, provided the required affidavit, enabling us to obtain visas for the United States.

It took several months for our visas to arrive and for our passage to be arranged. After leaving Austria, my mother, brother, and I lived in Zurich in a pension, *Comi ander Ekkehardstrasse* (Figure 3.8), which had been reserved before we left; it still exists today. We had two rooms, one for my parents, which was only occupied by my mother since my father had been retained in Vienna, and one for my brother and me. Robert and I enrolled in a neighborhood public school where we rapidly learned *Schwyzertütsch* (the Swiss German dialect). Speaking *Schwyzertütsch* was not only a way of belonging, but it also provided us with a secret language which our mother did not understand. The pension

Figure 3.7. The card table now at the Jewish Museum in Vienna

was located on a hill, and my brother and I had to walk down the hill to the school. Sometimes we went on roller skates, and one day, as I was going home, I slipped and fell on steps leading down the hill. I did not think I was badly hurt, but when I arrived at the pension, my mother was clearly horrified at the sight of my face covered with blood. Once she had cleaned off the blood, it turned out that I had only a small cut on my forehead, from which I still bear a scar. However, the image of my frightened mother stayed with me. I not only never wanted to roller skate again, but also I did not allow my son Mischa to have roller skates as a child.

When summer came, we left Zurich and went to La Baule, a beach resort on the Atlantic coast in Brittany, France, where our Uncle Ernst Papanek had established a summer colony for refugee children. The children were mainly from Jewish families, though there also were some whose parents were political refugees. During the late 1930s and early 1940s, Ernst organized a number of these children's homes. (He described these efforts in his autobiography [**Papanek, 1975**].)[2] Robert and I, along with our cousins Gus and George (Ernst's children), as well as newfound friends, spent a blissful summer in La Baule swimming, building

---

[2]Ernst's philosophy was that one could rely on the common sense and intelligence of children. A hallmark of the homes was that they were run as cooperatives, with the children's input playing a significant role.

Figure 3.8. The pension as it appears today

sand castles (Figures 3.9a and 3.9b), and railroad tracks with tunnels, etc., in the estate grounds with sand brought up from the beach by multiple trips. Food was in short supply, but we were reasonably well-nourished. I did not realize it then, but my mother and the other grownups were extremely worried about the future. Somehow they kept this from us and gave us a happy summer. Both Robert and I looked back on this period as a wonderful experience. Roberto Benigni's film *Life Is Beautiful* is emblematic of my memories of those days.

Figure 3.9. (a) Our cousins and my mother, brother, and I in La Baule, 1938. (b) Robert and I with Gus and George Papanek in La Baule, 1938

During the summer, visas for the four of us finally arrived and my mother booked passage on the *Ile de France*. My mother, Robert, and I were ready to leave for the United States. There had been no news from my father, as far as I was aware, but he turned up at Le Havre a few days before the *Ile de France* was scheduled to depart for New York. From my perspective, it was exactly what my mother had told me would happen: we would all go to America together. When my father joined us in Le Havre, Robert and I asked him what jail had been like. He told us that he had been treated well and cheerfully described how he had passed the time teaching the guards to play chess. One aspect of my father's personality, which strongly influenced both my brother and me, was to make something positive out of any experience.

For a long time we did not know details of my father's release from jail and, in fact, some details remain unclear. My father was kept in jail until he had signed over all our possessions, as mentioned above. The Austrian State Archive records that he was released in June 1938, but was allowed to leave Austria only in September 1938. Not only had my Uncle Edu posted a $5,000 bond for his release, my father had to pay several taxes to the Nazi regime to be allowed to leave Austria. After the war, a number of people (e.g., a jailer and an administrator involved with running the prisons) wrote us claiming credit for my father's release. There was no evidence as to who had done what, if anything. Nevertheless, in the years following the end of the war my parents sent CARE food packages to all the claimants and wished them well.[3] My parents also sent CARE packages to many other people we knew, such as Mitzi, who was not Jewish and had survived the war.

---

[3]CARE, Cooperative for Assistance and Relief Everywhere, was a humanitarian organization that distributed nonperishable food packages after the Second World War to help overcome food shortages in the war-torn countries.

Chapter 4

A New Life in America

————— ⚬╲╱⚬ —————

When we arrived in New York early in the morning on October 8, 1938, I stood on the deck looking for the Statue of Liberty, which I had read about. As it appeared out of the mist, it was very special for me and probably for all those entering the United States during the war (Figure 4.1). The symbolism associated with the Statue of Liberty may seem trite (and somewhat deceptive given our present immigration policies), but then it really was a welcoming sign for people who had been living in fear. Most of the immigration formalities had been taken care of by Uncle Edu (Figures 4.2 a and b, 4.3), so that a few hours after our arrival we boarded a train to Boston with him.

During our initial weeks in the United States, we were lodged in Brighton, a neighborhood of Greater Boston, where a large mansion had been transformed into a welcoming center for refugee families. We were taught about America, what it was like to live there, given lessons to improve our English, and aided in the steps required to be allowed to remain in the United States as refugees.

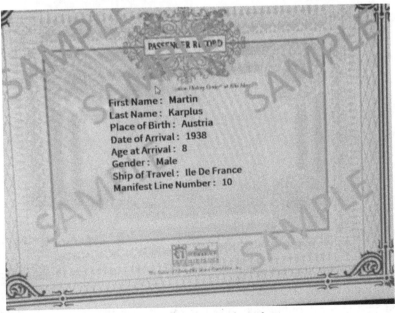

First Name: Martin
Last Name: Karplus
Place of Birth: Austria
Date of Arrival: 1938
Age at Arrival: 8
Gender: Male
Ship of Travel: Ile De France
Manifest Line Number: 10

Figure 4.1. Sample arrival certificate

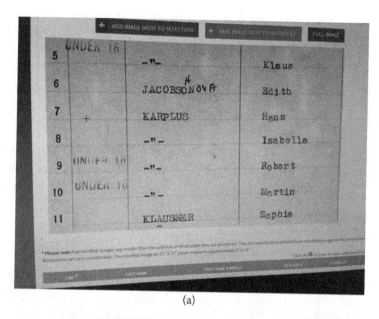

(a)

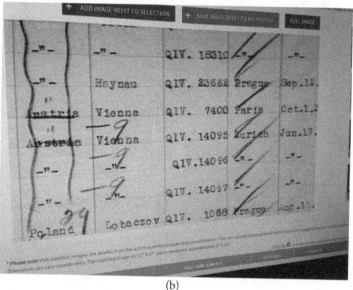

(b)

Figure 4.2. (a) Ellis Island certificate for my arrival. (b) Ellis Island documents, the arrival point, 1938

Soon we were ready to start a new life. My parents rented a small apartment in Brighton (Figure 4.4), and Bob, as he came to be called, and I were immediately enrolled in the local public schools. For me, it was the Harriet A. Baldwin Elementary School within walking distance of where we lived. I was in the third

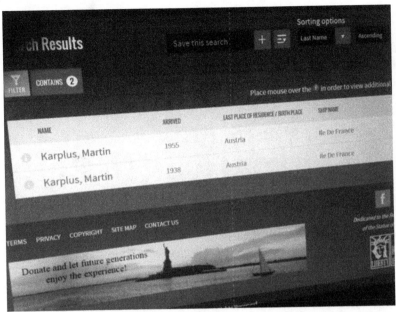

Figure 4.3. Ellis Island document showing that I also went on the Ile de France *when I returned from Europe in 1955*

grade and had the good fortune to have a teacher (I had a crush on her) who gave me special English lessons after school. At the age of eight, my English advanced so rapidly by being in school and by playing with the neighborhood children that these special lessons, alas, lasted only a few months.

I very much wanted to be accepted in my new country, and for a while I even refused to speak German at home, despite my parents' limited English. While living in Brighton, I was a street kid in every sense, hanging out with my friends, playing stickball and other such games, occasionally stealing candy just for the fun of it. One afternoon when I was playing in the street, I tripped in my haste to get out of a car's way and ended up with my foot under the car's rear tire. The driver had stopped and gotten out to see what was wrong, but with me screaming and crying it took him a while to understand that he should move the car off my foot. Once he did, thankfully my foot was only mildly sore. He wanted to drive me home and report to my parents what had happened, but I insisted that I was fine and would get home by myself. It was true that I was not hurt, but my primary concern was to keep my parents from knowing what had happened, how I spent my time, and specifically that I played in the streets.

Figure 4.4. Our apartment building in Brighton

My brother and I were friends with the two daughters of our family physician in Vienna, who now also lived in Brighton. We used to go to the local movie theater with them on Saturday afternoons. One of the serials,[1] *Hawk of Wilderness*, had as its hero a character like Tarzan. Every episode would end with the Hawk in a situation in which he surely could not survive; for example, going over a cliff. In fact, such film series were actually called "cliffhangers." We would wait breathlessly for the next Saturday's chapter, which usually backtracked slightly and had the hero catch a vine at the last second and swing to safety. An element of our foursome was that I had a crush on the younger sister, who had eyes only for my brother.

Our affluent life in Vienna was a thing of the past, and we were now relatively poor. My father had been able to get some money out of Austria by buying valuable stamps and putting them on envelopes, which he then mailed to relatives in the United States. In addition, he had purchased some stocks in the Netherlands, which were transferred to the United States before the Nazi invasion in 1940. This money enabled my parents to provide an affidavit for my maternal grandmother, Tante Mania as we called her, to come to the United States; her husband Onkel Samojla had died from a heart attack in 1939. (The expressions "Tante" and "Onkel" were used rather loosely to designate relatives.)

---

[1] *Hawk of the Wilderness* was a Republic Movie serial with twelve episodes, each about seventeen minutes in length. They were popular in Saturday matinees until the mid-1950s, when TV series began to replace them.

Despite our economic straits, my parents did everything to ensure that our lives were as unchanged as possible. The first summer we were in the United States, my parents worked as domestics—my father as a handyman and my mother as cook and cleaning woman—for a wealthy family who had an estate in Holderness, New Hampshire. We lived in a small house on the grounds, where Bob and I enjoyed an idyllic summer. The next summer my parents were similarly employed at a boys' camp, enabling Bob and me to spend the summer in the camp.

When the Wermer family, cousins of my mother, came to the United States in November 1939, about a year after we did, they had no resources and it took a while for Paul Wermer to be accredited as a physician. My parents invited their son Johnny to spend six months with us in Brighton. Johnny had been spoiled as an only child and was a rather picky eater when he joined our family. At meals my mother placed the food on the table and we were allowed to serve ourselves. Johnny soon discovered that Bob and I would gobble up everything unless he also dug in. According to his parents, Johnny's eating habits were completely transformed by his stay with us. He also learned to participate in the daily household chores, such as drying the dishes when my father washed them. We all enjoyed doing this because it gave us a chance to chat with him.

My parents enrolled in classes in order to find better employment. My father had been educated in physics at the University of Vienna and always had an interest in understanding how things worked. In 1939, he took a one-year course at Boston's Wentworth Institute. At that time Wentworth was not a degree-granting institution and mainly had students like my father who were learning a trade. It was mechanical engineering in my father's case. Although the United States was not yet engaged in the Second World War, it had begun supplying airplanes to Great Britain, generating a high demand for mechanical engineers. After completing his course at Wentworth, my father almost immediately obtained a job in Nonantum, a neighborhood within Newton, Massachusetts, at a company that made hydraulic airplane pumps. He started as a mechanic but rapidly rose to the position of inspector and worked at the company until his retirement. During those years, everyday he lunched at a nearby Italian restaurant, always having *Cotoletta alla Milanese*, that is, Wiener Schniztel. As a youth, I would occasionally join him there for lunch. My father frequently told stories to Bob and me about problems he had solved or how he had suggested improvements in the pumps. The way he described his ideas helped arouse our scientific curiosity.

My mother also went back to school. She first attended Simmons College, where she obtained a bachelor degree specializing in food science in 1952. She soon found a position as a dietician at the Beth Israel Hospital in Boston and rapidly advanced to be the chief dietician. Her position was similar to the one she had held at the Fango Heilanstalt in Vienna, but the staff she directed was much larger. While working, she studied part-time at Boston University and received a Master's degree in education at the age of 60 (Figure 4.5).

Motivated by their concern for our education, my parents decided to move to Newton (a suburb of Boston), where the schools were recognized as being superior to those in Boston.[2] They bought a small house (Figure 4.6) in a pleasant neighborhood in West Newton, where I enrolled in junior high school and where my brother transferred to Newton High School. The house was one of four that had recently been built by a developer at the top of a hill halfway between Newton High and my school, the Levi F. Warren Junior High School. We had little contact with our neighbors because our social life at the time involved primarily other refugee families from Austria, now living in Boston and nearby communities, such as Brookline (Figure 4.7).

Figure 4.5. My mother's MA degree from Boston University

---

[2] An exception was the Boston Latin School, where my brother had been accepted to enter in the seventh grade. It still is an outstanding school, which selects its students based on an entrance exam; it is the oldest existing high school in the United States, having been established in 1635.

Figure 4.6. Picture in 2016 of what had formerly been our house, 259 Otis Street, Newton, Massachusetts

Figure 4.7. Playing checkers with my mother

To my knowledge, we were the first Jewish family to live in Newton.[3] One Saturday, after we had lived in West Newton for about six months, an FBI agent knocked on the door and politely requested to see my father. The agent explained that he was investigating a complaint from our next-door neighbor, who had telephoned the FBI to report that every morning as he was leaving for work, my father would step out on the front porch, turn around, and make the Nazi salute while shouting "Heil Hitler." The FBI agent appeared rather embarrassed and said that he realized that such an accusation against a Jewish refugee from Nazi Austria was ridiculous. After some questioning regarding our family history and present legal status, he got up to leave and told us that nothing further would happen.

Figure 4.8. My father, Edu, and Walter Karplus in Belmont, Massachusetts, 1969

That refugees were not always welcome is also illustrated by my Uncle Edu's situation. He had a house with a large garden in an exclusive part of Belmont, a suburb of Boston, which at that time did not allow Jews to live there.[4] We were rarely invited to visit him when we first came to the United States and when we did, we had to hide our "Jewishness." We were not to say anything before entering the house to avoid the neighbors' overhearing of our foreign accents. As time went on, Edu and his wife Harriet became more relaxed about their neighbors and numerous reunions were held at their home involving various Karplus brothers (Figure 4.8) as well as cousins and extended family from Austria.

---

[3]The Historical Atlas of Massachusetts (University of Massachusetts, Amherst, 1991) provides a map with a description of the migrations of Boston's Jewish community. German and Eastern European Jews settled in downtown Boston. It was only in the 1960s that many Jewish families became members of the middle class and began to populate suburbs like Brookline and Newton, among others. Uncle Edu's move to Belmont in the 1930s was clearly an exception.

[4]At that time such restrictive covenants, concerned with Jews as well as other minorities, were common throughout the northeastern United States. Real estate developers, homeowners, and neighborhood associations made these restrictions. They are often referred to as "red-lining," because restricted areas were, and still are, shown by red lines on local maps. Such restrictions have generally faded away for Jews, but still exist for African Americans in many neighborhoods.

My junior high teachers soon realized that I was bored with the regular curriculum, so they let me sit in the back of the classroom and study on my own. What made this experience particularly nice was that another student, a very pretty girl named Martha Palmer, was given the same privilege, and we worked together. The arrangement was that we could learn at our own pace without being responsible for the day-to-day material, but we did have to take the important exams. Several dedicated teachers at Warren Junior High helped us when questions arose, particularly with science and mathematics. With this freedom, we explored whatever interested us and did much more work than we would have done if we were only concerned with passing the required subjects.

While in junior high school, I worked for a while at a local food market, which delivered to customers. My job was to put orders into boxes for the deliveries. The manager soon realized that I was very able and put me in charge of the delivery team. I had to do spot checks to make sure that the other people in the group, some of whom were older the I was, filled the orders correctly. I still had an accent at the time and there was some resentment of my position by some of the other delivery "boys."

As part of our education, we had to take several technical courses. The two I chose were printing and home economics, the latter because the students did real cooking. I had become interested in cooking early on and used to spend time in the kitchen of West Newton with my mother and grandmother. My mother cooked simply but well, while my grandmother and I helped (see Chapter 19). The final exam in the cooking course had us prepare a dinner for the class, with each group responsible for one dish. In nonacademic activities, I participated like everyone else although I was not particularly good in sports. Importantly, I made a number of friends, with whom I formed a close-knit group through high school.

Chapter 5
Beginning of Scientific Interests

—⊸❦⊹—

Soon after we moved to Newton, our parents gave Bob a chemistry set, which he augmented with materials from the school laboratory and the drug store. He spent many hours in the basement generating the usual bad smells and making explosives. I was fascinated by his experiments and wanted to participate, but he informed me that I was too young for such dangerous scientific research. My plea for a chemistry set of my own was vetoed by my parents because they felt that this might not be a good combination—two teenage boys generating explosives could be explosive! Instead, my father had the idea of giving me a Bausch & Lomb microscope (Figure 5.1).

Initially I was disappointed—no noise, no bad smells—although I soon produced the latter with the infusions I cultured from marshes, sidewalk drains, and other sources of microscopic life. I came to treasure this microscope, and more than

Figure 5.1. Looking through the microscope

sixty years later it is still in my possession. An especially rewarding aspect of my working with the microscope was that my father, who was a thoughtful observer of nature, spent a lot of time with me and was always ready to come and look when I had discovered something new. In some ways these interactions compensated for the fact that I had been jealous of my brother when we were little because he would often climb into bed with my father in the morning to work on some mathematical problems. I would sit on the floor next to the bed listening but, of course, the problems were generally beyond me.

I had found an exciting new world and passed many hours looking through my microscope. The first time I saw a group of rotifers, I was so excited by my discovery that I refused to leave them, not even taking time out for meals (Figure 5.2).

They seemed the most amazing creatures to me, as they swam across the microscope field with their miniature "rotary motors," which were called cilia, which they used to steer in search of food. My enthusiasm was sufficiently contagious that some of my friends came over to look through the microscope and view the rotifers. This was the beginning of my interest in nature, which developed into my passion for science. It was nurtured by my father and encouraged by my mother, even though it was still assumed that I would go to medical school and become a doctor.

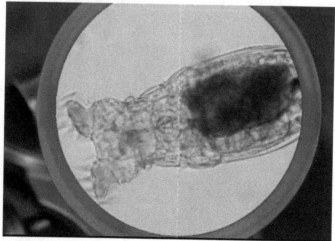

Figure 5.2. Image of a rotifer as seen through a microscope

One day my closest friend, Alan MacAdam, saw an announcement of the Lowell Lecture Series.[1] The series that had caught Alan's eye was entitled "Birds and Their Identification in the Field," to be given in January 1944, by Ludlow Griscom, the curator of ornithology at the Museum of Comparative Zoology at Harvard University. Alan and I had occasionally walked in the green areas in Newton, particularly the Newton Cemetery, and looked for birds with my father's old pair of binoculars, a relic of his service in the First World War. I was enthralled by Griscom's first lecture, which provided insights into bird behavior and described the large number of species one could observe within a fifty mile radius of Boston. It amazed me that it was possible to identify a given species from "field marks," even from only a glimpse of a bird, if one knew how and where to look. Alan did not attend the subsequent lectures, but I continued through the entire course. At the end of the fourth or fifth lecture, Griscom came up to me and asked me about myself. He then invited me to join his field trips, and a new passion was born (Figure 5.3).

My treasured microscope was relegated to a closet, and I devoted my free time to observing birds on my own, as well as with Griscom and his colleagues, with the Audubon Society, and other groups that organized field trips. It was a wonderful experience for me to participate in field trips with expert "birders" like the Argues and Rosario Mazeo, the concert master of the Boston Symphony Orchestra at the time. I also joined the spring walks of the Brookline Bird Club in the Mt. Auburn Cemetery in Cambridge. Many birds, particularly colorful warblers, use it as a rest stop in their migration from

Figure 5.3. As a birder with my father's binoculars on a field trip

---

[1] A Boston institution, founded in 1836 by John Lowell, Jr., a Boston "Brahmin" as the moneyed Boston upperclass was traditionally called. He left funds to a trust for "the maintenance and support of public lectures" that were to be free and open to all regardless of gender or race. They were organized as evening courses on a wide range of subjects at the Boston Public Library. Excellent lecturers were invited from the many universities in the Boston area, as well as from nonacademic institutions. The lectures still exist; in 2016–17, for example, they focused on Shakespeare, commemorating the 400th anniversary of his death.

the Caribbean to their nesting places in the northern United States and Canada. The Brookline Bird Club was founded in 1913 and is still active today, more than a century later.

The culmination of these trips was the annual "census," usually held at the height of the bird migration in May. This was an activity sponsored by the Audubon Society, and the objective was to observe (see or hear) the largest number of species within a given 24-hour period. Each year, Griscom organized one such field trip, inviting only a select group of "birders" to participate. The census lasted a full 24 hours, starting just after midnight to find owls in the woods and rails and other aquatic species in the swamps. There was a carefully planned route, based on known habitats and recent sightings of rare species. The specific itinerary was worked out in a meeting during which everyone contributed the interesting birds they had sighted recently, but Griscom made the final decision on how the census would proceed. As the youngest (by far) in the group, I was assigned special tasks. One such task—perhaps not the most pleasant—was to wade into the swamp at night (fortunately there was a moon) and scare up birds so that they would fly off and could be identified by their calls. We did not have to see the birds to add them to the list. Another task, which recognized my abilities at bird identification, was to "shadow" one of the other birders who had aroused Griscom's suspicions. The man would frequently go off by himself and report that he had seen a rare bird, which sometimes would not be found by anyone else. Probably for that reason, a rule was made by Griscom that at least two people had to be involved in any observations that were included. On this census we found 160 or so different species, a record at the time for the area.

I became intrigued by alcids, of which the now extinct flightless great auk was the most spectacular member of the family. I persuaded my parents to go for summer vacation to the Gaspé Peninsula in Canada where a famous rocky island, Perzé Rock, just offshore had nesting colonies of two kinds of alcids: razorbilled auks and guillemots (Figure 5.4).

Although one could not visit the island (it had been made a sanctuary in 1919 and access was forbidden to protect the nesting birds), I had borrowed a telescope from the Audubon Society to view the birds. We drove through the Gaspé Peninsula by car, spending nights in bed-and-breakfast places where the owners spoke both English and French. One strong impression of the trip through New

1947   The Bulletin of the Massachusetts Audubon Society   21

Massachusetts Alcids

By Martin Karplus*

ALFRED O. GROSS

Razor-billed Auks and Brunnich's Murres, two of the alcids that reach New England shores, are here shown on their rocky breeding grounds farther north.

Figure 5.4. My first scientific publication

Brunswick and the Gaspé area was that the houses in most of the villages were in much poorer condition than those on the coast, which were supported by the tourist trade. Moreover, the small villages were often dominated by an outsized church, indicating the power of religion in these communities.

We stayed several days in an inn near the shore and I spent most of the day looking through the telescope at the nesting birds. Watching the very beautiful and spectacular gannets, as well as razorbilled auks, and the occasional puffin, was one of my most exciting birding experiences.

Many of the alcids that nest in the Gaspé and farther north go south for the winter to New England where they usually are far out to sea. However, in the event of a storm they are likely to be blown close to shore, so that such storms (particularly nor'easters) present the best opportunity to see rare species, such as the tiny little auk, which is only eight inches in length but can survive in the roughest ocean. For several days in the winter of 1944, school was closed because of a heavy snowstorm and I took the early morning train up to Gloucester, a town on Cape Ann, north of Boston, and hiked out to the shore. I sat on the steps of one of the large mansions, shuttered for the winter. It provided an excellent position from which to survey the ocean. The day started well as I quickly spotted several alcids in the cove. As I was getting chilled after a couple of hours and ready to walk back to the train station, a car pulled up. A couple of men got out and walked slowly toward me. At first, I naively presumed they were other birders, interested in what I had seen. I soon realized that they did not look like birders—no binoculars for one, and not really dressed for a day in the snow. Moreover, they approached me in a rather aggressive manner, asking me what I was doing and why. They did not believe that I was sitting in such a storm looking for birds. Shortly after, they showed me their police badges and bundled me into the car, not for a ride to the train station, but instead to the Gloucester police station. I was only fourteen years old and very frightened and became even more so when I realized from their questions that they thought I might be a German spy. One has to remember that this happened in 1944 when the Second World War was still going on. That I was an immigrant from Austria who spoke German and had German-made Zeiss binoculars did not help. They suspected me of signaling to German submarines ("U-boats") off the coast, preparing to land saboteurs or whatever.[2] It took hours of interrogation and several phone calls to the Audubon Society until someone who knew me answered. The officers finally decided that I was not doing anything wrong and drove me to the train station. That was the last time I ventured on such a trip by myself.

One project that I helped to organize under Griscom in 1945 was a field study of an area in Wayland around Hurd's Pond [**Griscom, 1949**], a varied terrain where over 200 different species had been observed. To make the survey, some members of the group visited the area every weekend from February through June and recorded the birds that they saw or heard. In addition, records were made of nesting pairs to compare with earlier surveys. Of the latter, some, like the indigo bunting, no longer nested there, while others, like the catbird, had significantly

---

[2] In May 1945, when the Second World War with Germany had ended, a number of U-boats, which had been patrolling the east coast, surrendered to US Naval Forces.

increased due to the fact that there were more open areas. It would be interesting to know what the populations are today, more than seventy years later.

On one of the field trips to Newburyport with Griscom, I spotted an unusual gull. When I pointed it out to him, he concluded, after looking through his telescope, that he had never seen a bird like that and that we should try to "collect" it, a euphemism for shooting the bird. He had a license to carry a "collecting gun," which had a pistol grip and a long barrel that made it easier to aim. The bird was far away and separated from us by mud flats, which were only partly exposed at low tide. I was given the task of collecting the bird even though I had never fired a gun other than at fairs. I waded out fairly close to the bird and successfully shot it. After a careful comparison with birds in the Museum of Comparative Zoology collection, Griscom was convinced that we had found something new, a hybrid between a Bonaparte's gull (common in America) and a European black-headed gull (common in Europe but rare in North America), which had somehow crossed the ocean (Figure 5.5).

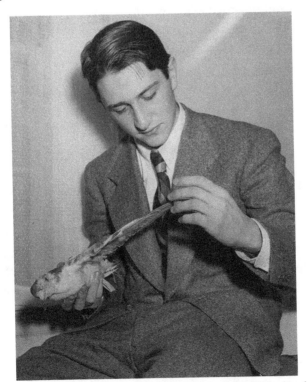

Figure 5.5. With the presumed hybrid gull, showing its wing feathers, which played an essential role in its identification

For many years the gull was in Harvard's Museum of Comparative Zoology's bird collection with its wing feathers beautifully spread out. It is now stored in the Peabody Essex Museum in Salem with other birds from Essex County.[3]

I entered Newton High School in the fall of 1944 and soon discovered that I did not have the same supportive environment as in elementary and junior high school. My brother, Bob, had graduated from Newton High School two years before and had done exceedingly well. My teachers presumed that I could not measure up to the standards set by my brother. Since I had always striven to keep up with Bob, this just reinforced my feelings of inferiority. Particularly unpleasant were my interactions with the chemistry teacher. When my brother suggested I compete in the Westinghouse Science Talent Search, the chemistry teacher, who was in charge of organizing such applications, told me that it was a waste of time for me to enter and that it was really too bad that Bob had not tried. However, I talked to the high school principal and he gave me permission to go ahead with the application. I managed to obtain all the necessary forms without encouragement from anyone in the school. A test was part of the selection process and I found a teacher who was willing to serve as a proctor. I did well enough to be invited as one of the forty finalists to Washington, DC. Each finalist had a science project for exhibition in the Statler Hotel, where we were staying. My project was on the lives of alcids, based in part on the trip to the Gaspé Peninsula and some of the field studies I had made during the New England winters. The various judges spent considerable time talking to us, and the astronomer Harlow Shapley, who was the chief judge, charmed me with his apparent interest in my project. I was chosen as one of two co-winners. At that time, there was one male winner and one female winner: Rada Demereck and I were co-winners. The visit to Washington, DC was a great experience, especially because we met President Truman (Figures 5.6 and 5.7), who welcomed us as the future leaders of America. Moreover, winning the Westinghouse Talent Search made up for the discouraging interactions with some of my high school teachers. Their attitude contrasted with that of my fellow classmates, who voted me "most likely to succeed."

---

[3] It would now be possible to test this identification by means of its DNA. For many species of gulls, "bar codes" were developed to identify the gulls that were sucked into the jet engines of airplanes. Although I have tried to get genetic experts to test the DNA available from the specimen, none have agreed to undertake the analysis.

1947

President Harry Truman with the finalists at the White House

Figure 5.6. President Truman with the forty Westinghouse Science Talent Search finalists in 1947

My final forays into ornithology took place during several summers at the end of high school and after I entered college. In 1947, I had a summer internship at the Maryland Patuxent Research Refuge of the Fish and Wildlife Service, the only National Wildlife Refuge established to conduct research (Figure 5.8).[4]

Publicity about the harmful effect of DDT on bird life had prompted studies at the refuge. We collected eggshells as part of a field survey of two several-acre plots: one sprayed with DDT at the normal level and the other without DDT. My task was to analyze eggshells for their DDT contents and to determine the differences between shells (their thickness and other features) from the two plots. I found that the eggshells of birds in the areas with DDT were thinner and more likely to break. My observation probably disappeared in some government file.[5] The summer was an exciting one for me and a fine introduction to field research and laboratory work as part of a team. It was very hot, and all my colleagues drank beer to relax in the late afternoon. I did not like the bitter taste initially but soon learned to enjoy beer, particularly when it was ice cold.

---

[4] The name is an Algonquin word meaning "water running over loose stones."

[5] Almost 25 years later in June 1972, as a result of a lawsuit brought by the Environmental Defense Fund, a nationwide ban on the use on DDT as a pesticide was signed. An argument for the ban was that the osprey population in the eastern United States had been endangered by DDT, which had been found to be in high concentration in their eggs. They made a rapid comeback once the ban was put into effect, and one can see this magnificent bird now nesting on telephone poles along the New England coast.

THE BOSTON HERALD, WEDNESDAY, MARCH 5, 1

Newton High Senior Wins $2400 Prize

WASHINGTON, March 4—Martin Karplus, 16, of 259 Otis street, West Newton, Mass., a refugee with his family from Austria, tonight was awarded one of the two $2400 top scholarships in the Westinghouse Science Talent Search in which 3200 high school seniors competed.

The other winner was Miss Vera Radoslava Demerec, 16, daughter of Dr. Milislav Demerec, director of the department of genetics at Carnegie Institution, Cold Spring Harbor, N. Y. The awards were announced at a banquet here tonight.

Karplus, a senior at Newton High School, is at 16 one of the leading ornithologists in the country, having contributed to scientific magazines from his knowledge of birds. His goal is to solve one of nature's great mysteries—the bird's uncanny orientation ability.

The Massachusetts winner came to this country in 1938, when his family fled from Vienna a few days after Hitler invaded Austria. His parents are Mr. and Mrs. Hans Karplus. His brother, Robert, 19, is a freshman adviser at Harvard, where he is teaching and studying for his doctor's degree.

Winners of $400 four-year schol-

MARTIN KARPLUS

arships included Paul Leroy Cloke, 18, 49 Forest avenue, Orono, Me., and Irene Elizabeth Nagy, 17, 134 Andover street, Bridgeport, Ct. Winners of $100 one-year awards included James McKenna, 17, 3 Union street, Lebanon, N. H., and Clarence Leslie Gregory, Jr., 16, 14 Maher avenue, Greenwich, Ct.

ST. ALBANS ELECTS FISHER

ST. ALBANS, Vt., March 4—Former Mayor B. W. Fisher, Republican, was elected mayor today.

Figure 5.7. Boston Herald article about the Science Talent Search

Thanks to a meeting with Professor Robert Galambos, who did research on the echolocation of bats (His laboratory was in the basement of Memorial Hall at Harvard), I was invited by his collaborator, Professor Donald Griffin, to join his group in a study of bird orientation that was to take place in Alaska during the summer of 1948.[6] Our team was based at the Arctic Research Laboratory in the town of Point Barrow, Alaska, which is located at latitude 71° north, the northernmost point of land on the North American continent. The laboratory was run by

---

[6]Griffin and Galambos were well known because they had demonstrated in 1940 that bats used echolocation to orient themselves and find their prey. Lazzaro Spallanzani had actually suggested this in 1794, but apparently it was not generally accepted.

Figure 5.8. Sign for the Patuxent Refuge

the Office of Naval Research,[7] primarily to study how soldiers and sailors adapt to life in the Arctic winter. Its director, Lawrence Irving from Swarthmore College, however, had a broad view of the laboratory's mission and had invited our group to use the facilities (Figure 5.9).

Our primary interest was in golden plovers, which nested on the tundra in areas not far from the laboratory. Atlantic and Pacific golden plovers nested together, and the two species separated in the fall to migrate over a thousand miles southward to their respective winter homes. We trapped some plovers of both species and attached radio transmitters to all of them and magnets to half of them. We then released the plovers twenty to fifty miles from their nests, in both the Atlantic and Pacific directions, and followed them at a distance with a small airplane. The idea was to ascertain if the birds with magnets would have a more difficult time returning to their nesting area. There were suggestive results: the birds with magnets seemed to get more disoriented than those without, though they all found their way back to their nesting area. However, Griffin felt that what we had found was not conclusive proof that the plovers had a magnetic sense. I was disappointed that this meant that we were not going to publish our results.

---

[7]It is worth mentioning that before there were organizations like the National Science Foundation (NSF) to support civilian research, ONR was the primary governmental source of funding. This was due primarily to Vannevar Bush, who headed the Office of Scientific Research and Development during the Second World War. Bush realized the importance of basic research to the future of the United States, and it was through his initiative that the ONR funded basic research until the NSF was established in 1950.

Figure 5.9. The Arctic Research laboratory staff in front of the building with me in the foreground

Griffin's rigor had a strong effect on me and contributed to my own concern about publishing unless the results were clearly valid.

In Umiat, an observation camp located 170 miles from the main laboratory, I organized an experiment with the aid of the other scientists, who must have been amused by my youthful enthusiasm. (I was seventeen years old and by far the youngest member of the research team.) The experiment involved several nesting pairs of robins. Three people participated in the observation of the nests: each person took a four-hour shift twice a day to have information for the full 24 hours of daylight in the Arctic summer. We found that the robins fed the offspring over the entire 24-hour period and, interestingly, that the young robins left the nest earlier than did their cousins, who nested in Massachusetts. I wrote a paper [**Karplus, 1952**] describing the results, with the conclusion that the survival value of the shorter time in the nests, which were highly exposed to local predators, made up for the dangers of the longer flights required to reach the summer nesting area in the Arctic. It is still not clear whether my conclusion was correct, although it did stimulate a number of papers, both pro and con, and the paper continues to be cited [**Schekkerman et al., 2003**]. As an aside, I noticed that at Umiat, where we were on our own, the normal 24-hour day stretched to about 30 hours; we stayed awake for 22 or so hours and then slept for about eight.

The following summer Griffin invited me to Cornell University, where he was on the faculty before he joined the Biology Department at Harvard. In addition to conducting experiments in the Griffin lab, I enjoyed "hanging out" with college students and school teachers who were taking summer courses at Cornell. I initially worked on bat echolocation and was much impressed by the way the bats were kept in the refrigerator between experiments; they went to sleep, hanging from a rod all in a row. At Griffin's suggestion, I then focused on trying to condition pigeons to respond to a magnetic field to test the results of an article that had concluded that pigeons use the earth's magnetic field to navigate. I was doubtful about the paper because I thought the analysis was flawed. My attempts at conditioning the pigeons did not succeed, and we never published our negative (to me, positive) results. Subsequently, other experiments have shown that pigeons, as well as wild birds, do use the earth's magnetic field as an aid in navigation. This experience taught me that being skeptical is essential in science, but that it is also important to be receptive to new ideas even if you do not like them.

Chapter 6

College Years

———————◎⊱◎————————

I entered Harvard in the fall of 1947. There was never any question about my wanting to attend Harvard and I did not apply to any other school. My parents were concerned that I might not be admitted (Harvard was reputed to limit the admission of Jewish students, though apparently there was no quota at any time), and so they asked Richard von Mises, a Harvard professor who had received his doctorate in Vienna and come to the US as a refugee, to write a recommendation letter. He interviewed me for about thirty minutes and then agreed to write a letter, as he had done for my brother. In addition to the Westinghouse scholarship, I received a National Scholarship from Harvard to cover the cost of living on campus (Figure 6.1).

Without it, I would have had to live at home to save money. I would not have minded since I was not a rebellious teenager eager for independence and distance from my parents. However, as I soon discovered, much of the Harvard experience took place on campus outside of classes, at dinner and in the evening.

When I entered Harvard, I still intended to go to medical school but I changed my mind during the freshman year. When I told my parents, they fortunately were supportive about this change in plans. I always felt that whatever my choice for my future, they would be there for me, to help in any way they could. In a way, this was a continuation of the family environment of my childhood in Vienna.

My teenage ornithological studies, fostered by Griscom and Griffin, had already introduced me to the fascinating world of research, where one is trying to discover something that no one has ever known. I began to think about doing research in biology, but I had concluded that to approach biology at a fundamental level ("to understand life"), a solid background in chemistry, physics, and mathematics was imperative. I enrolled in the Program in Chemistry and Physics, which was unique to Harvard at the time. It exposed undergraduates to courses in both areas at a depth that they would not have had from either one alone. There was also the additional advantage from my point of view that it was less structured than chemistry. For example, it did not require *Analytical Chemistry*, certainly a good course as taught by Professor James J. Lingane, but one that did not appeal to me. Although I shopped around for advanced science courses to meet the rather

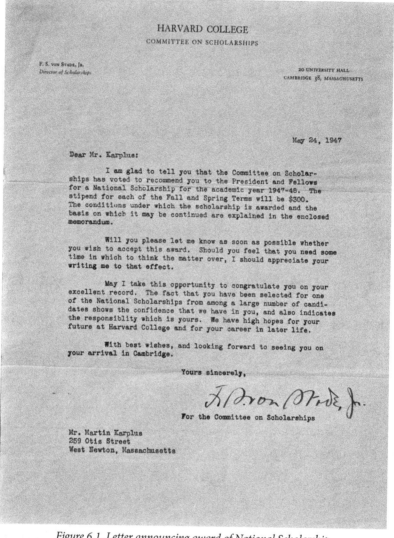

Figure 6.1. Letter announcing award of National Scholarship

loose requirements, I also enrolled in *Elementary Chemistry* because it was taught by Leonard Nash. A relatively new member of the Harvard faculty, Nash had the reputation of being a superb teacher. Elementary chemistry in Nash's lectures was an exciting subject. A group of us (including DeWitt Goodman, Gary Felsenfeld, and John Kaplan, who later became a law professor at Stanford) had the special privilege that Nash spent extra time discussing with us a wide range of chemical questions, far beyond those addressed in the course. The interactions in our group, though we were highly competitive at exam times, were also supportive. This freshman experience confirmed my interest in research and the decision not to go to medical school.

Harvard provided me with a highly stimulating environment as an undergraduate. Its *laissez-faire* policy allowed one to take any course with the instructor's permission, even without having the formal prerequisites. The undergraduate dean said it was up to me to decide and, if a course turned out to be too much for me, that would be "my" problem. I enrolled in a wide range of courses, chosen partly because of the subject matter and partly because of the outstanding reputation of the lecturers. These courses included *Democracy and Government* given by Louis Hartz, a young professor who had the reputation of being a charismatic teacher. A great teacher he was indeed, though because of my socialist leanings, I did not always agree with him. Another course was *Abnormal Psychology* by Robert White, whose lectures on the origin of personality disorders were fascinating. More related to my long-term interests were some advanced biology courses, in which I registered without having to suffer through elementary biology and biochemistry. Two memorable courses were George Wald's *Molecular Basis of Life* and Kenneth Thimann's *Plant Physiology* with its emphasis on the chemistry and physiology of growth hormones (auxins). Both professors were inspiring lecturers and imbued me with an excitement for the subject. The courses emphasized that biological phenomena ("life itself") could be understood at a molecular level, which has been a leitmotif of my subsequent research career. Wald's course also introduced me to the mechanism of vision, which led to my first paper on a theoretical approach to a biological problem [**Honig and Karplus, 1971**].

Although my undergraduate career at Harvard was a formative experience and reinforced my interest in science, it was reminiscent of my high school days in that my brother had preceded me, had been a stellar student in the Program in Chemistry and Physics, and was in the process of completing a PhD at Harvard with E. Bright Wilson, Jr., and Julian Schwinger. I spent considerable time with Bob and his fellow graduate students in Wilson's group. I was tolerated, I suppose, as Bob's little brother, though one day I made my mark when I solved a problem (they were always "challenging" each other) before any of them did. Given the importance of this "success" in my life, I am fond of the problem and shall restate it here: "It is agreed that to divide a pie between two people so that both are satisfied, one is allowed to cut the pie in two and the other chooses. The problem is how to extend this concept (dividing or choosing) to three or more people, so that everyone is satisfied." There is a special solution for three people and a general solution for any number.[1] After solving this problem, I was accepted as part of the group by my brother and his friends, and during the afternoons when I could escape from the many labs I had to do, I would often join them. Their discussions of science exposed me to new ideas that I would not have come across otherwise.

The legendary *Elementary Organic* course taught by Louis Fieser was a standard part of the Program in Chemistry and Physics. I was convinced it would be a waste of time because it had the reputation of requiring a very tedious laboratory and endless memorization. An early version of what became a well-known textbook by Louis and Mary Fieser was available in lecture-note form and Bob had a copy of it. Rather than enrolling in the course, I tried to learn organic chemistry by reading it on my own. In retrospect, I have realized that if I had "suffered" through the course, it would have provided me with a better knowledge of organic reactions that would have been useful for my later research focus on biological systems.[2]

After studying the Fieser lecture notes, I enrolled in Paul Bartlett's *Advanced Organic Chemistry*, which taught the physical basis of organic reactions. It was an excellent course, though difficult for me because one was supposed to know many organic reactions, which I had to learn as we went along. At one point, Bartlett suggested in class that we read Linus Pauling's *The Nature of the Chemical Bond*, which had been published in 1939 based on Pauling's Baker Lectures at Cornell. The *Nature of the Chemical Bond* presented chemistry for the first time as an integrated subject that could be understood, albeit not quite derived, from its quantum chemical basis. The many insights in this book were a critical element in orienting my subsequent research.

---

[1] I will leave finding a solution to the reader or to looking it up on the Web, where several possible solutions can now be found.

[2] An interesting sidenote concerns me being invited to give a plenary lecture for the Organization for the Prohibition of Chemical Weapons (OPCW) Day that took place on May 2, 2016, in celebration of the ninth anniversay of the acceptance of the Chemical Weapons Convention by the United Nations. My invitation stemmed from the coincidence that OPCW received the Nobel Peace Prize in 2013, the same year that I received the Chemistry Prize. In preparing my lecture, I made much use of the Web, particularly Wikipedia, because I knew very little about chemical weapons. Among other items, I learned that Fieser had invented napalm, a type of jellied gasoline, in a secret laboratory at Harvard in the early 1940s; napalm continued to be used in warfare by the United States and other troops with horrible effect in the Vietnam and Korean Wars, and even in the Second World War. It is forbidden by a UN Convention to be used against civilians, but not for military objectives.

At the end of three years at Harvard, I needed only one more course to complete the requirements for a bachelor degree. During the previous year, I had done research with Ruth Hubbard and her husband, George Wald.[3] I mostly worked with Hubbard rather than Wald, because I was interested in the chemistry of vision and she had a deeper knowledge of the chemical properties of retinal, the visual chromophore. When I brought up my need to find one additional course to be able to graduate in three years, Wald suggested that I enroll in the physiology course at the Marine Biological Laboratory in Woods Hole, Massachusetts. This course was one of the few non-Harvard courses that was accepted for an undergraduate degree by the Faculty of Arts and Sciences. The physiology course was widely known as a stimulating course designed for postdoctoral fellows and junior faculty. The lectures were presented by scientists who were doing research in the laboratory while summering in Woods Hole. It offered students a state-of-the-art view of biology and biological chemistry.[4] For me, the only undergraduate in the course, it was a wonderful experience. I not only learned a great deal of biology and biochemistry, but I also met several people, including Jack Strominger and Alex Rich, who became lifelong friends.

Woods Hole was an exciting place. Among the famous scientists I met there was Otto Loewi, who had received a Nobel Prize in Physiology and Medicine (1936) for the discovery of the chemical basis of the transmission of nerve impulses. He conducted an experiment that had much in common conceptually with the demonstration by my paternal grandfather that the hypothalamus used hormones to communicate with the body, rather than nerve impulses. Loewi isolated a frog heart, stimulated it by exciting the vagnus nerve, took some of the fluid in which the heart had been bathed and exposed another heart to it; as a result the second heart contracted. Perhaps my grandfather did not receive a Nobel Prize because the work was done in the early 20th century (the key paper was published in 1909) when the Nobel Prizes were just beginning; the first prizes were awarded in 1901. Another Nobel Laureate I met was Albert Szent Györgyi. His prize in

---

[3] Although Hubbard was scientifically on par with Wald, she remained a senior research associate, a nonprofessorial appointment, until very late in her career when she was finally "promoted" to professor. This was not an uncommon fate for women in science at the time.

[4] The course, which was first taught in 1892, still exists and its subject matter changes from year to year to keep up with new developments in physiology and molecular biology. It now provides an intensive training at the interface between cellular and computational biology. When I took the course, "computational biology" did not exist. I think it is fair to say that my research has contributed to making it a central part of biology.

Physiology and Medicine (1937) was awarded for discovering vitamin C and showing that it existed in high concentration in paprika, a staple of the Hungarian diet. He wrote a book *The Nature of Life, A Study of Muscle* [**Szent-Györgyi, 1948**]. This little book as well as discussions with him inspired my interest in biological research. Another such book was Schroedinger's *What is Life: The Physical Aspects of the Living Cell* [**Schroedinger, 1944**], which presented a physicist's view of biology. Not everything in these books has turned out to be correct, but what was important for me was their presentation of the logical development of science in general, and of an approach to understanding living systems in particular.

Many of the famous scientists summering in Woods Hole held court in the afternoons at the nearby beach and fascinated us would-be scientists with their discussions of new experiments and scientific gossip. In addition, there was an active student life, since many of the senior researchers brought along students from their labs.

Woods Hole provided a wide range of marine animals, which were caught for use in the research that was done there, as well as for the laboratory experiments in the physiology course. A prime example was the squid, whose giant axon was the ideal system for studies of the mechanism of nerve conduction. Since most of the experiments used only a small fraction of the animal, each week or so I collected some of the leftover laboratory squids and lobsters and prepared a feast for myself and friends, who supplied bread, wine, and salad.

Chapter 7

Graduate School

⚜

In considering graduate school during my last year at Harvard, I had decided that I wanted to go to the west coast because I was interested in seeing other parts of the United States. This would offer me the opportunity to visit natural wonders such as the Grand Canyon and Yosemite National Park. I applied to the Chemistry Department of the University of California at Berkeley and to the Biology Department of the California Institute of Technology (Caltech). Accepted at both, I found it difficult to choose between them. Providentially, I visited my brother Bob in the spring of 1950 at the Institute of Advanced Studies in Princeton, New Jersey. He was working there as a postdoctoral fellow with J. R. Oppenheimer. Bob, who was always looking out for me and my future, introduced me to Oppenheimer. Einstein happened to walk by while Bob was showing me around and he introduced me to Einstein as well.[1] Given Einstein's fame, I am not sure I would have been able to think of something to say after we were introduced. Fortunately, he immediately inquired about my interests, to which I replied that I hoped to study biology from a chemist's viewpoint. He gave me a warm smile, said, "That's good," with a German accent, and shook my hand. I remember Einstein as a large man, mainly because of his big halo of white hair, though actually he was not that tall.

When Oppenheimer asked me about my plans, I told him of my dilemma in choosing between U. C. Berkeley and Caltech for graduate school in chemistry or biology. He had held simultaneous appointments at both institutions and strongly recommended Caltech, describing it as "a shining light in a sea of darkness." His comment influenced me to choose Caltech, and I discovered that Oppenheimer's characterization of the local environment was all too true. Pasadena itself held little attraction for a student at that time. However, camping trips in the nearby desert and mountains and the vicinity of Hollywood made up for what Pasadena lacked.

I had become very interested in movies and shortly after I arrived at Caltech I organized a series showing classic films. In part, this was to make up for the

---

[1] Einstein had left Germany in 1933 and joined the Institute, which had recently been established. He continued to work there until he died in 1955.

lack of things to do in Pasadena and, in part, it enabled me to watch many films that I had always wanted to see. The series showed mainly silent films accompanied by live piano music played by fellow Caltech students. (One of them was Walter Hamilton, a crystallographer, who also came to Oxford a year after I did.) In searching for films, I gained access to several production studios, where I would ask the librarians to lend me films for our nonprofit Caltech series. A high point was my visit to the Chaplin Studio. The receptionist was not particularly forthcoming, but then Charlie Chaplin himself walked in. Chaplin was one of my cinematic heroes at the time, so I recognized him immediately. To my amazement, he asked who I was and what I wanted. He seemed intrigued by the idea that a science student was interested in films. I asked him about the possibility of showing *Monsieur Verdoux*, which I had not seen. He said it had been withdrawn for political reasons; apparently it was poorly received in part for its dark humor and in part for Chaplin's "left wing" views. Chaplin told the librarian to let me have some of his early short films, which at the time also were not available to the public. I remember this meeting as one of the very special events of my graduate career. It was some years until *Monsieur Verdoux* was screened again and I had a chance to see it.

At Caltech, I first joined the group of Max Delbrück in biology. He had started out as a physicist but, following the advice of Niels Bohr, had switched to biology. With Salvador Luria and others, he had been instrumental in transforming phage genetics into a quantitative discipline. His research fascinated me, and I thought that working with such a person would be a perfect entrée for me to do graduate work in biology. Many bright and lively people were working with Delbrück, including Seymour Benzer who, like Delbrück, was a former physicist. Seymour and I became friends and we had many discussions of phage genetics, biology, as well as a variety of other subjects of mutual interest, including the relative advantages of horsemeat and beef. I had not known that horsemeat was available in the United States, though I had read it was eaten in France. Because horse filet was very inexpensive, it became a staple in our household diet later in my graduate career.

After I had been in the Delbrück group for a couple of months, he proposed that I present a seminar on a possible area of research. I intended to discuss my ideas for a theory of vision (how the excitation of retinal by light could lead to a nerve impulse), which I had started to develop while doing undergraduate research with Hubbard and Wald. Among those who came to my talk was Richard Feynman. I had invited him to the seminar because I was taking his quantum mechanics course and knew he was interested in biology, as well as everything else. I began the seminar confidently by describing what was known about vision but was interrupted after a few minutes by Delbrück, who was sitting in the back of the room, with the comment, "I do not understand this." The implication of his remark, of course, was that I was not being clear, and this left me with no choice but to go over the material again. As this pattern repeated itself (Delbrück saying "I do not understand" and me trying to explain), I had not even finished my ten-minute introduction after thirty minutes and was becoming very nervous. When Delbrück intervened yet again, Feynman, who was also sitting in the back of the room several seats away from Delbrück, turned to him and whispered loud enough so that everyone could hear, "I can understand, Max, it is perfectly clear to me." With that, Delbrück got red in the face and rushed out of the room, bringing the seminar to an abrupt end. Later that afternoon, Delbrück called me into his office to tell me that I had given the worst seminar he had ever heard. I was devastated by this and it was agreed that I could not continue to work with him. It was only years later that I learned from reading a book dedicated to him that what I had gone through was a standard rite of passage for his students—each one gave the "worst seminar he had ever heard." One consequence of this "rite" was that many of the students who did work with him had more self-confidence than I had at the time.

After the devastating exchange with Delbrück, I spoke with George Beadle, chairman of the Biology Department, who suggested that I find someone else in the department with whom to do graduate research. However, I felt that I wanted to go "home" and transferred to the Chemistry Department. I joined the group of John Kirkwood, who was doing research on charge fluctuations in proteins as well as on the fundamental aspects of statistical mechanics and its applications. I undertook work on proteins and the research started out well. It was complemented by a project involving Irwin Oppenheim and Alex Rich. Kirkwood's course *Advanced Thermodynamics* was famous for its rigor, and the three of us, with Kirkwood's encouragement, worked together to prepare a set of lecture notes for the course. Each of us was responsible for writing up some of the lectures and the other two read them over. This was very useful for our learning thermodynamics and the set of notes was circulated widely. Some years later,

Irwin Oppenheim prepared an improved version of the notes and it was published as a text entitled *Chemical Thermodynamics* [**Kirkwood and Oppenheim, 1961**]. I must admit that I was surprised that Alex and I were not listed as coauthors, but that did not alter our lifelong friendship with Irwin, and, in fact, the three of us lived for fifty years within walking distance of each other.

In the spring of 1951, as I was getting immersed in my research project, Kirkwood received an offer from Yale. Linus Pauling, who was no longer taking graduate students, asked each student who was working with Kirkwood whether he would like to stay at Caltech and work with him (Figure 7.1).

I was the only one to accept; all of Kirkwood's other students, who had worked with him for a longer time, decided to go to Yale, Irwin among them. In retrospect, I think I made an excellent decision. I was to be Pauling's last graduate student.[2] Initially, I was rather overwhelmed by Pauling. Each day upon arriving at the lab, I found a hand-written note on a yellow piece of paper in my mailbox, which always began with something like, "It would be interesting to look at ...." As a new

Figure 7.1. Linus Pauling with molecular models

[2]The Linus Pauling Award was established by the Oregon Local Section of the American Chemical Society for outstanding contributions to chemistry. The first award was given to Pauling himself in 1966, and he invited me to give a lecture in a symposium organized in his honor. I received the award in 2004. The ceremony was held at Oregon State University in Corvallis, Oregon, where my nephew Andy Karplus is on the faculty. After dinner he read part of a letter, which he found in the Pauling Archive. I had written to Pauling, who had invited me to participate in a party celebrating his 85th birthday. It happens that his birthday is on February 28, the same day as that of my son Mischa, and I wrote him apologizing for not attending his birthday party with the sentence: "Nobody will notice my absence at your celebration, but Mischa would be very unhappy not to have me here." In looking through the file of correspondence between Pauling and myself, I came across several recommendation letters that he had written for me. In one, to Professor Mayer at UCSD he wrote, "I am pleased to learn ... that you are trying to persuade Martin Karplus to accept an appointment at UCSD.... I think he is the most able theoretical student to have studied with me."

student I took this as an order and tried to read all about the problem and work on it, only to receive another note the next day, beginning in the same way. When I raised this concern with Alex Rich and other postdocs, they laughed, pointing out that everyone received such notes and that the best thing to do was to file them or throw them away. Pauling had so many ideas that he could not work on all of them. He would communicate them to one or another of his students, but he did not expect a response. Once I knew that, my relationship with Pauling became more constructive.

Given Pauling's interest in hydrogen bonding in peptides and proteins, he proposed that I study the different contributions to hydrogen bonding interactions for a biologically relevant system, but I felt this would be too difficult to do in a rigorous way. Because quantum mechanical calculations for molecules still had to be done with calculating machines and tables of integrals (something hard to imagine when even log tables have followed dinosaurs into oblivion), we had to find a simple enough system to be studied by quantum mechanical theory. I chose the bifluoride ion (FHF^-) because the hydrogen bond is the strongest known, the system is symmetric, and only two heavy atoms are involved. (Such "strong" hydrogen bonds have become popular as factors contributing to enzyme catalysis, although there is no convincing evidence for their role.)

I sometimes felt intimidated when I went in to talk with Pauling, but it was wonderful to work with him and be exposed to (although not necessarily understand) his intuitive approach to chemical problems. One day I asked Pauling about the structure of a certain hydrogen bonded system (i.e., whether the hydrogen bond would be symmetric, as in FHF^-). He paused, thought for a while (he always sat back in his chair and looked up at the ceiling), and gave a prediction for the structure. When I asked him why, he thought again and offered an explanation. I left his office and soon realized, in part by discussing it with my circle of friends, that his explanation made no sense. So I went in to ask Pauling again, thinking that perhaps he would come up with a different conclusion. Instead, Pauling said that he believed that the predicted structure was correct, but he proposed an entirely different explanation. After going over his analysis, I was again dissatisfied with his rationale and caught up with him as he was leaving his office. He produced yet another rationale. This one made sense to me, so I did some crude calculations that indicated that Pauling was indeed correct. What amazed me then, and still does today, is that Pauling came to the correct conclusion, apparently based on intuition, without having worked through the analysis. He "knew" the right answer, even if it took more thought to figure out why.

Research under Pauling's guidance was very rewarding, all the more so because of the intellectual and social atmosphere of the Chemistry Department at Caltech. Professors—like Verner Schomaker and Norman Davidson, as well as Pauling—treated graduate students and postdoctoral fellows as equals. We participated in many joint activities including trips into the desert. Also, we held frequent parties at our Altadena house, and Feynman would occasionally come and play the drums (Figure 7.2).

Figure 7.2. Feynman playing the bongo drums

At one such party, Pauling disappeared for a while and I discovered him out in the backyard on his knees collecting snails, which had infested our yard, for his wife Ava Helen to cook for dinner. (It was only later in France, when I collected my own snails, that I learned how complicated it was to prepare them—the snails had to fast for a week—so that now, looking back, I am not sure what the Paulings did with their snails.) The Paulings often held court at their Altadena house with a swimming pool, which had been constructed for Pauling to exercise after he had lost a kidney. These visits were special, not only because of the opportunity to speak with Pauling in such a relaxed atmosphere, but also because he had a son Peter and a lovely daughter Linda close to my age.

We generally served wine at the parties in our house. Because we could not afford good wine for many people, we would usually buy some inexpensive jug wine and serve it in "recycled" bottles of good wine. It was amusing, to say the least, to have our guests sip these wines and comment on their characteristics. Since then, I have always been somewhat suspicious of self-proclaimed wine connoisseurs. One of the "experimental" chemistry efforts for these parties was to obtain pure ethyl alcohol from the lab, dilute it appropriately, add caramelized sugar and other ingredients to create "Bourbon whiskey," which was also much appreciated by many of our guests.

Pauling's presence at Caltech attracted many postdoctoral fellows. As a graduate student, I was the "baby" of the group, which included Alex Rich, Jack Dunitz, Massimo Simonetta, Leslie Orgel, Edgar Heilbrunner, and Paul Schatz. Interacting with them was a wonderful part of my Caltech education and many of them became my friends. Massimo Simonetta, in particular, remained a dear friend and colleague after his return to Italy and I visited him and his wife Miriella regularly in Milan, as well as on ski trips to Courmayeur in the winter and to Portofino in the summer.

My parents had given me their old car as a graduation present, and several times during my Caltech career I drove across the country to our home in Newton, Massachusetts, for part of the summer. Each time I took a different route, once through Canada with visits to the Banff and Jasper National Parks, and another time through the Deep South. On one such trip, while driving through Texas on a very hot summer day, my friends and I decided to take a swim and cool off. We passed one designated swimming area that was full of people. Because we were unshaven and dirty from the long drive, we looked along the river for a quieter place to swim. About a mile downstream we came upon another swimming area with broken-down steps leading to the water. As it was deserted, we decided it was the perfect place for us. After we had been in the water for about ten minutes, a couple of pickup trucks drove up and several men jumped out with guns at the ready. They turned out to be the local law enforcement officers; they ordered us out of the water and demanded to know what we thought we were doing—"white folks swimming in an area reserved for n--." They had noticed the Massachusetts license plate on the car and had concluded we were northern "troublemakers." After some effort we succeeded in explaining that, given our scruffy state, we had just not wanted to bother other (white) people. The officers let us go, with the admonition that we had better drive straight through Texas without stopping anywhere, which we did.

Graduate students in chemistry had to take an oral qualifying exam as part of their preparation for research. I had little trouble responding to the questions posed by the members of the committee until Pauling asked me to describe his theory of metals. I had studied this and confidently began by saying, "Let us consider copper." "So what is the atomic number of Cu?" Pauling chimed in. After a moment of silence, he continued, "Well, start with hydrogen and helium and go through the periodic table, and when you get to copper you will know its atomic number." It rapidly became clear that I was not going to be able to do this, and Norman Davidson, a member of the committee, took pity on me and told me the atomic number. After a few more questions on which I did fine, although I remained somewhat flustered, Pauling got up and congratulated me on passing the exam, but he said I should come into his office. There he gently admonished me, saying that every chemist should know the periodic table and that I should be ready to recite it in the future. I did learn it and could recite it for many years, but have never been asked about it again.

My initial attempt at a purely *ab initio* approach to the bifluoride ion failed and I soon realized that it was necessary to introduce experimental information concerning the atomic states to obtain a meaningful estimate of the relative contribution of covalent and ionic structures. I developed a method for doing this and completed my work only to discover that William Moffitt had just published a similar approach called the "methods of atoms in molecules."[3] He had presented the method in a more general and elegant formulation [**Moffitt, 1954**], while my treatment focused specifically on the bifluoride ion. Although there were significant differences in the details of the methodology, I felt so discouraged by the similarities that I never published "My Great Idea," as Verner Schomaker, one of the members of my thesis committee, called it. Not too surprisingly, I also found it difficult to begin writing my thesis. Nevertheless, I retained my interest in this type of approach, and some years later Gabriel Balint-Kurti joined my group as a graduate student and we proposed an improved version of the theory, which we called the Orthogonalized Moffitt method and applied it to the potential energy surfaces for simple reactions [**Balint-Kurti and Karplus, 1969**]. Such calculations are mostly of historical interest today, when fast computers and *ab initio* programs are widely used without empirical corrections. Large systems, like protein and nucleic acids, are an important exception because empirical potential functions still are necessary to permit calculations of their internal motions, as we will see later.

---

[3]Moffitt, who was to be my predecessor at the Harvard Chemistry Department, died from a heart attack during a squash match at the age of 33 in 1958. For several years after that the department felt there was no theoretical chemist good enough to replace him and so only appointed nontenured assistant professors to take care of teaching the required courses. It was only in 1966 that I was appointed professor as his replacement.

Chapter 8

Postdoctoral Sojourn in Oxford and Europe

———— ⟨❧⟩ ————

One day in October 1953, Pauling came into the office I shared with several postdocs and announced that he was leaving in three weeks for a six-month trip and that "it would be nice" if I finished my thesis and had my exam before he left. This was eminently reasonable, since I had finished the calculations some months before and I had received a National Science Foundation (NSF) postdoctoral fellowship to go to England that fall. Pauling's "request" provided just the push I needed, even though the introduction to my thesis was all I had written thus far. With so much to get done, I essentially wrote day and night, with my Caltech colleagues typing and correcting what I wrote. In this way, the thesis was finished within three weeks, and I was able to have my final PhD exam and celebratory party before Pauling left. I went back to Newton, Massachusetts, for a visit with my parents, before going to New York where I boarded a ship, the *Ile de France*, for England and arrived shortly before Christmas 1953.

Why had I chosen to go to Europe as a postdoctoral fellow? Although I did not think of it at the time, in retrospect I feel that I had a desire to return to, or at least to visit Europe, where I was born. Given my traumatic departure from Austria in 1938 and the continued anti-Semitism there, I looked in other countries to find a research group in line with my interests.

Oxford and Cambridge, England, were two of the best centers for theoretical chemistry in Europe. In my NSF postdoctoral application, I had proposed to work with John Lennard-Jones in Cambridge, England. He, however, had left Cambridge to become Principal of Keele University in 1953, and so I had to alter my plans. I decided to join Charles Coulson at Oxford University, where he had an active group in theoretical chemistry at the Mathematical Institute. There were regular seminars with lively questioning by which I deepened my knowledge of quantum chemistry. At Caltech, there had been much less of this and, although I had many friends, in my thesis research I had worked mainly on my own. One member of the Coulson group was Simon Altmann, who greatly improved my limited knowledge of group theory and who, with his wife, Bochia, "adopted" me while I was in Oxford. In addition, there were visitors such as Don Hornig and Bill Lipscomb, who were there on sabbatical leaves.

I was 23 years old when I arrived in England. Having worked continuously all the way through graduate school, I was eager to have the sojourn in Europe provide experiences beyond science. The NSF postdoctoral fellowship consisted of what was then a generous salary of $3,000 per year, sufficient to do considerable traveling. I took the NSF guidelines about following the customs of the institution quite literally, perhaps more so than was intended, and traveled throughout Europe outside of the three eight-week terms when I was in residence in Oxford. For many of the Oxford undergraduates, the eight-week terms were an almost continuous party time, while they did their actual studying between terms.

Oxford University was (and still is) an umbrella organization made up of a set of colleges, which play an essential role for the faculty, made up of college fellows, and for undergraduates, who live in the colleges and meet with a fellow on a weekly basis to report on their progress. This is somewhat less true in the sciences than the humanities because of laboratory work in the former. I was aware of this aspect of Oxford and had applied to be a visiting member of Balliol College, known as one of the most intellectual colleges. In response to my application (which had included a support letter from Pauling and Coulson), I received a one-sentence reply from the Master: "We already have our quota of foreigners for the year." When I mentioned this to Coulson, he made me a member of his college, Wadhouse. I was so put off by what had happened with Balliol[1] that I did not take advantage of being a member and never participated in any of the college activities.

I arrived in Oxford shortly before Christmas and promptly went to see Coulson to greet him and, more importantly, to have him sign my NSF form, to be able to start my fellowship. Then I left almost immediately on the first of many trips to Paris, where I met up with Sidney Bernhard, a friend from Caltech days. He had arrived in Cambridge, England, two years before and was helpful in introducing me to Paris, its arts, architecture, and culture, and most importantly its many

---

[1] I was appointed Eastman Visiting Professor for the academic year 1999–2000. This prestigious position had been founded by George Eastman, the head of Eastman Kodak. One aspect of being Eastman Professor is that one is automatically a visiting fellow of Balliol College. At the reception in my honor held at the college when I arrived, I must admit I enjoyed recalling my earlier experience.

excellent and relatively inexpensive restaurants. In those days, even most modest restaurant had good, sometimes simple, classical French cuisine. Unfortunately, this is no longer true. The Restaurant des Beaux Arts across the street from the École des Beaux Arts was such a place; although it still exists, it is now a "tourist trap."

An expensive restaurant we did manage to visit was Lapérouse, one of the best restaurants in Paris with three Michelin stars at the time (see Chapter 19). I had the idea of going there as a reporter (I had a card from the *Harvard Crimson*) with the object of seeing what it was like in such a restaurant for a foreigner who knows little French. Monsieur Topolinsky, the proprietor, welcomed Sidney and me, showed us around, and invited us to a wonderful dinner in one of their private rooms (Figure 8.1).

Before sitting down to dine, he and the *maître d'hôtel* took us into the cellar, which held several thousand bottles, some very old and expensive. A special rack contained bottles whose labels had been washed away in one of the floods of the Seine. Monsieur Topolinsky and the *maître d'hôtel* would open a bottle and then discuss the vintage (some they concluded were over 50 years old), the chateau the wine came from, and its specific characteristics (Figure 8.2).

Sidney and I participated in the tasting, but of course had nothing to contribute to the discussion. Even now that I know much more about wines, I still would not be able to do what they did. Although I did not write anything for the *Crimson*, the whole experience seemed so special that I wrote an article for the *New Yorker* and sent it to A. J. Libling, whose articles about food and France I had appreciated for a long time. He wrote thanking me for the story, but ultimately it was not accepted for publication.

During the two years in Oxford as a postdoctoral fellow, I spent more time thinking about chemical problems than actually solving them. My aim was to find areas where theory could make a contribution of general utility in chemistry. I did not want to do research the results of which were of interest only to theoretical chemists. From the literature, listening to lectures, and talking with scientists like Don Hornig and the Oxford physicist H.M.C. Pryce, I realized that magnetic resonance was a vital new area, with chemical applications of magnetic resonance in their infancy. It seemed to me that nuclear magnetic resonance (NMR), in particular, was a field where theory could make a contribution. Chemical

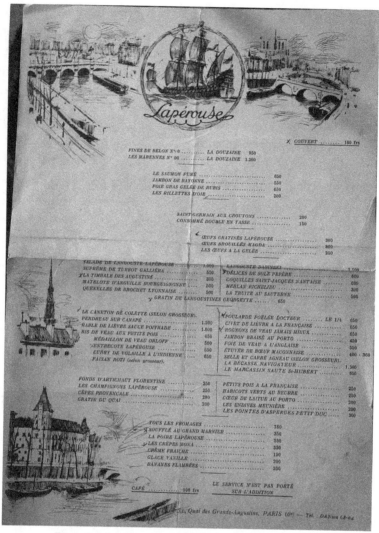

Figure 8.1. Lapérouse menu marked with the dishes we ate

shifts, for example, could provide a means of testing theoretical calculations, but, even more important, quantum mechanical theory could aid in interpreting the available experimental results and propose new applications.

My first paper in chemistry on the quadrupole moment of the hydrogen molecule, obtained from different approximate wave functions, was an example [**Karplus, 1956**]. Although this was a short paper, I spent an enormous amount of time

Figure 8.2. M. Topolinsky and the maitre d' *in the cellar*

rewriting and polishing it before finally submitting it. When students seem to face similar problems in finishing their first paper, I often tell them about what I went through and assure them that publishing (or being ready to publish) one's work becomes easier with each succeeding paper.

Chapter 9

University of Illinois: NMR and Coupling Constants

———— ◉✦◉ ————

A s my postdoctoral fellowship in Oxford (1953–1955) neared its end, I decided to return to the United States to try to begin my academic career there. With my growing interest in magnetic resonance, I focused on finding an institution that had active experimental programs in the area. One of the best schools from this point of view was the University of Illinois in Urbana-Champaign, where Charles Slichter in Physics and Herbert Gutowsky in Chemistry were doing pioneering work in applying nuclear magnetic resonance (NMR) to chemical problems. The University of Illinois had a number of openings in the Chemistry Department at that time because it was undergoing a radical renovation. Several professors, including Roger Adams, who had been chairman for nearly thirty years, had retired. Pauling recommended me to the University of Illinois and the new chairman Herbert Carter offered me a job without an interview and without waiting for a recommendation from Coulson, which presumably was delayed by having been mailed from overseas. The latter was fortunate, because Coulson had written that, although he had no doubt about my intellectual abilities, I had done very little work on problems he had suggested. This was true; they seemed of limited interest to me. I accepted the offer from Illinois without visiting the department, something unimaginable today with the extended courtships that have become an inherent part of the academic hiring process. The University of Illinois offered me an instructorship at a salary of $5,000 per year. The offer did not mention start-up funds, unlike present-day offers, and I did not think of asking for research support. In my case, the latter would have involved a desk calculator as the most expensive item.

Although the University of Illinois was a very good institution with excellent chemistry and physics departments, it was in a small town in the flat rural Midwest where I could not imagine living for more than five years. Having had such a good time as a postdoctoral fellow traveling in Europe, I was ready to get to work, and Urbana-Champaign was a place where I could concentrate on science with few distractions. The presence of four new instructors[1]—Rolf Herber, Aron Kupperman, Robert Ruben, and I—plus other young scientists on the faculty, such as Doug Applequist, Lynn Belford, and E. J. Corey, led to a very interactive and congenial atmosphere.

I focused a major part of my research on theoretical methods for relating nuclear and electron spin magnetic resonance parameters to the electronic structure of molecules. The first major problem I examined was concerned with proton-proton coupling constants, which were known to be dominated by the Fermi contact interaction. What made coupling constants of particular interest to me was that for protons, which were not bonded, the existence of a nonzero value indicated that there was an interaction beyond that expected from localized bonds. In the valence bond framework, which I used in part because of my training with Pauling, nonzero coupling constants provide a direct measure of the deviation from the perfect-pairing approximation. To translate this qualitative idea into a quantitative model, I chose to treat the vicinal coupling constant in a molecule like ethyl alcohol, one of the first molecules to have its NMR spectrum analyzed experimentally. Specifically, I studied the proton (H)–proton (H′) coupling in the HCC′H′ fragment as a function of the HCC′H′ dihedral angle (Figure 9.1), a relatively simple system consisting of six electrons with neglect of the inner shells. I believed that it could be described with sufficient accuracy for the problem at hand by including only five covalent valence-bond structures. To calculate the contributions of the various structures, I introduced semiempirical values of the required molecular integrals. The calculations of the vicinal coupling constant for a series of dihedral angles were time consuming and it seemed worthwhile to develop a computer program. This was not as obvious in 1958 as it is now. Fortunately, the ILLIAC, a "large" digital computer, had recently been built at the University of Illinois. If I remember correctly, it had 1,000 words of memory, which was enough to store my program. The actual program was written by punching holes in a paper tape. If one made a mistake, the incorrect holes were filled with nail polish. The most valuable aspect of having a program for such a simple calculation, which could have been done on a desk

---

[1] The four of us, Herber, Kupperman, Ruben, and I, were Jewish. This constituted a sharp break with the appointments under the chairmanship of Roger Adams, who was known for being anti-Semitic. There were no Jewish faculty members in chemistry during his tenure.

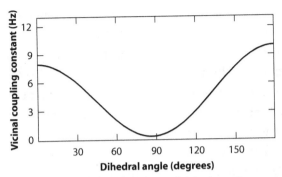

Figure 9.1. Dihedral angle dependence of vicinal coupling constant

calculator, was that once the program was known to be correct, a large number of calculations could be performed without having to worry about arithmetic mistakes.

Just as I finished the analysis of the vicinal coupling constants [**Karplus, 1959a**], I heard a lecture by R.U. Lemieux on the conformations of acetylated sugars. I do not remember why I went to the talk. It was an organic chemistry lecture, and the chemistry department at Illinois was rigidly separated into divisions (organic chemistry, physical chemistry, etc.), which had a semiautonomous existence. Lemieux reported measurements of vicinal coupling constants and noted that there appeared to be a dihedral angle dependence, although the details of the behavior were not clear. The results were exciting to me because the experiments confirmed my theory, at least qualitatively, before it was even published.

E. J. Corey, then an assistant professor at Illinois and later a colleague at Harvard, was one of the people with whom I had dinner on a fairly regular basis at the Tea Garden, a passable Chinese restaurant in Urbana-Champaign. We usually discussed recent work of mutual interest and I described the studies that I had made of vicinal coupling constants. Corey immediately recognized the possibility of using the results for structure determinations and published what is probably the first application of my results in organic chemistry [**Bradshaw et al., 1959**].

Not long after, the theory was described in a comprehensive review of the use of NMR in organic structure determinations [**Conroy, 1960**], and someone introduced the name "Karplus equation" for the relationship I had developed. This proved a mixed blessing. Many people attempted to apply the equation to determine dihedral angles in organic compounds. They found some deviations of the measured coupling constants from the predicted values for known structures and published their results, commenting on the inaccuracy of the theory. As happens too frequently with the application of theoretical results in chemistry, most people who used the so-called "Karplus equation" did not read the original paper [**Karplus, 1959a**] and thus did not know the limitations of the theory. They assumed that because the equation had been used to estimate vicinal dihedral angles, the theory said that the coupling constant depends *only* on the dihedral angle. By 1963, having realized organic chemists tend to primarily write and read Communications to the *Journal of the American Chemical Society*, I published such a Communication [**Karplus, 1963**]. In it, I described various factors, other than the dihedral angle, that are expected to affect the value of the vicinal coupling constant: they include the electronegativity of substituents, the valence angles of the protons (HCC′ and CC′H′), and bond lengths. The main point of the paper was not to provide a more accurate equation but rather to make clear that caution had to be used in applying the equation to structural problems. My closing sentence, which has often been quoted, was the following: "Certainly with our present knowledge, the person who attempts to estimate dihedral angles to an accuracy of one or two degrees does so at his own peril."

Despite my concerns about the limitations of the model, the use of the equation has continued, and the original paper [**Karplus, 1959a**] is one of *Current Contents* "most-cited papers in chemistry". Correspondingly, the 1963 Communication was listed as one of the most-cited papers in the *Journal of the American Chemical Society* [**Dalton, 2003**]. In addition, there have been many empirical "extensions" of the equation; perhaps the most complex published form [**Imai and Osawa, 1989**] uses a twelve-term expression. Equations have been developed for vicinal coupling constants involving a variety of nuclei [**Karplus and Karplus, 1972**] (e.g., ^{13}C-CC′-H, ^{15}N-CC′-H), and they have been applied in areas ranging from inorganic to organic to biochemistry. An important more recent application is the use of these relationships as part of the data employed in structure determination of proteins by NMR [**Kline, Braun, and Wüthrich, 1988; Mierke, Huber, and Kessler, 1994**]. The vicinal coupling constant model, which was developed primarily to understand deviations from perfect pairing, has been much more useful than I would have guessed. I wrote in the *Encyclopedia of Magnetic Resonance*

[**Karplus, 1996**], "In many ways my feeling about the uses and refinements of the "Karplus equation" is that of a proud father. I am very pleased to see all the nice things that the equation can do, but it is clear to me that it has grown up and now is living its own life." I continued to work on problems in NMR and Electron Spin Resonance (ESR) because new areas of chemistry were being studied by these spectroscopic methods and it seemed worthwhile to provide insights from theoretical analyses of these applications. Examples are a study of the hyperfine interactions in the ESR spectrum of the methyl radical [**Karplus, 1959b**] and the contributions of π-electron delocalization to the NMR coupling constants in conjugated molecules [**Karplus, 1960a**]. My general approach to the magnetic properties of molecules was summarized in an article entitled "Weak Interactions in Molecular Quantum Mechanics" [**Karplus, 1960b**]. The choice of title was apt because the energies involved in coupling constants and hyperfine interactions are indeed weak relative to the electron volts involved in bond energies, excitation energies, and ionization potentials that are the bread and butter of quantum chemistry. However, the title also had a facetious aspect in that my brother had been working on what physicists call "weak interactions" [**Karplus and Kroll, 1950**].

At Illinois, my officemate was Aron Kuppermann. Our appointment at Illinois was the first academic position for both of us, and we discussed science, as well as politics and culture, for hours on end. We became fast and life-long friends. He and his wife, Roza, lived in an apartment next to mine and often invited me for dinner. Our friendship continued for more than fifty years, even after I left Illinois to go to Columbia University and Aron moved to Caltech. Having Aron and Roza as friends provided a special continuity in my life. Each time we met, our conversation would begin as if we had seen each other the day before.

When we were at the University of Illinois, Aron and I decided that, although we were on the faculty, we wanted to continue to learn and would teach each other. I taught Aron about molecular electronic structure theory and we published two joint papers on molecular integrals [**Karplus, Kuppermann, and Isaacson, 1958; Kuppermann, Karplus, and Isaacson, 1959**]. Aron in turn taught me about chemical kinetics, his primary area of research. Aron was officially an experimentalist, but he was also an excellent theoretician, as demonstrated by his landmark quantum mechanical study of the $H+H_2$ exchange reaction with

George Schatz. This work was some years in the future [**Schatz and Kuppermann, 1977**], but in the late 1950s we both felt that it was time to go beyond descriptions of reactions in terms of the Arrhenius formulation based on the activation energy and pre-exponential factor. My research in this area had to wait until I moved to Columbia University, where I would have access to the required computer facilities.

Chapter 10

Move to Columbia and Focus on Reaction Kinetics

———————— ❧ ————————

During the summer of 1960, I participated in an NSF program at Tufts University, whose purpose was to expose high school and college science teachers to what scientists actively engaged in research were doing. Our task was to present modern chemical concepts in a way that would help the teachers in the classroom. Ben Dailey, a Columbia professor and an organizer of the program, asked me one day as we were standing next to each other in the washroom whether I would consider joining the chemistry faculty at Columbia University. Because I had already been at Illinois for four of the five years I had planned to stay there, I responded positively. I heard from Columbia shortly thereafter and received an offer to join the IBM Watson Scientific Laboratory with an adjunct associate professorship at Columbia.

The Watson Scientific Laboratory was an unusual institution to be financed by a company like IBM. Although the Laboratory played a role in the development of IBM computers, many of the scientists there were doing fundamental research. The Lab was founded in 1945 near the end of the Second World War to provide computing facilities needed by the Allies. Its director, Wallace Eckart, is perhaps best known for his highly accurate perturbation calculation of the three-body problem posed by the motion of the earth around the sun in the presence of the moon; the $H + H_2$ reaction, which I studied while at the Lab, is also a three-body problem in the Born–Oppenheimer approximation. When Eckart described the position at the Watson Lab to me, he made clear that staff members were judged by their peers for what they did in their research and not for their contribution to IBM. The presence of outstanding scientists on the staff, such as Erwin Hahn, Seymour Koenig, Alfred Redfield, and L. H. Thomas (of Thomas–Fermi fame), supported this description and made the place very attractive. Moreover, the Watson Lab had a special advantage for me in that it had an IBM 650, an early digital computer, which was much more useful than the ILLIAC because of its greater speed, larger memory, and simpler (card) input. (No more nail polish!) I was to have access to considerable amounts of time on the IBM 650 and to receive support for postdocs, as well as other advantages over a regular Columbia faculty appointment. This was a seductive offer, but I hesitated about accepting a position that, in any way, depended on a company, even a large and stable one like IBM. This was based, in part, on my political outlook, but even more so on the fact that industry has as its primary objective making a profit, and all the rest is secondary. By contrast, my primary focus was on research and teaching, which are the essential aspects of a university, but not of industry. Consequently, I replied to Columbia and the Watson Lab that the offer was very appealing, but that I would consider it only if it included a tenured position in the Chemistry Department. Columbia acceded to my request and after some further negotiation, I accepted the position for the fall of 1960.

The environment at the Watson Lab was indeed fruitful, in terms of both discussions with other staff members and the available computer facilities. I was able to do some research there that would have been almost impossible to do at Columbia. However, not unexpectedly, the atmosphere gradually changed over the years. Richard Garwin, a brilliant physicist who had developed the design of the first hydrogen bomb with Edward Teller, assumed an increasing role at the Watson Lab and introduced pressure from IBM to do something useful (i.e., profitable) for the company, such as conducting meetings with people at the much larger and more applied IBM laboratory in Yorktown Heights. This would have been the equivalent to doing internal consulting. I decided in 1963 that the time had come to leave the Watson Lab and moved to the full-time professorial position in chemistry at Columbia. (IBM closed the Watson Lab in 1970.) Given that experience, I always warn my students and postdocs about accepting jobs in industry. They may well have an exciting environment when they first join the staff, receive a significantly higher salary than they would at a university, and they would not have to worry about obtaining grants to support their research. However, I urge them to remember, a new management team can take over at any time and decide to cut down on the research budget, which in the short term only costs money. This attitude has led to lay-offs of individual scientists or the closing of beautifully equipped research laboratories that were built only a few years before. It is of primary importance that your objectives (in my case, teaching and research) be the same as those of the institution where you work. This requirement is ideally satisfied in a good university but cannot be guaranteed in industry.

I continued research in the area of magnetic resonance after moving to New York. One reward of being at Columbia was the stimulation provided by interactions with new colleagues, such as George Fraenkel, Ben Dailey, Rich Bersohn, and Ron Breslow.[1] Frequent discussions with them helped to broaden my view of chemistry. In particular, my interest in ESR was rekindled by George Fraenkel and we published several papers together [**Karplus and Fraenkel, 1961; Karplus, Lawler, and Fraenkel, 1965; Lawler et al., 1967**], including a pioneering calculation of ^{13}C hyperfine splittings [**Karplus and Fraenkel, 1961**].

Although the techniques we used were rather crude, the results did provide insights concerning the molecular electronic structure and aided in understanding the measurements. Many of the weak interactions were a real challenge to estimate in the mid-1960s. Now they can be calculated essentially by pushing a button with programs like the widely used Gaussian package. The high-level *ab initio* treatments that are used routinely today have the drawback that, even though the results can be accurate, the insights obtained by the earlier, simplified approaches are often lost in the complexity of the calculation. A striking example of this is provided by the studies I did to analyze the vicinal coupling constants. The simple valence bond model I used showed how the perfect pairing approximation broke down. This information is lost in the multi-configuration calculations used today. Moreover, the new generation of researchers raised on this type of calculation often are not even interested in anything beyond obtaining the accurate number, which in many cases is already available from experiment.

My interest in chemical reaction dynamics had deepened at Illinois through many discussions with Aron Kuppermann, but I began to do research in the area only after moving to Columbia. There were several reasons for this. There is no point in undertaking a problem if the methodology and means for solving it are not available. It is important to feel that a problem is ripe for solution. (This has been a guiding rule for much of my research—there are many exciting and important problems, but only when one feels that they are ready to be solved should one invest the time to work on them. This rule has turned out to be even more relevant in the application of theory to biology, as we shall see later.)

---

[1]One day Ben Dailey asked me to be an invited speaker at a symposium on magnetic resonance he was organizing at an upcoming national meeting of the American Chemical Society (ACS). When he submitted the list of invited speakers, he was informed that only members were allowed to speak, even if they had been invited. This excluded me since I was not a member. At the time the ACS was mainly identified with chemical industry and, given my attitude about working for industry, I had decided not to become a member. Dailey felt it important to have me be a participant, so he persuaded the ACS to change the rule to make it possible for a non-member to present an invited lecture. Since that time, the ACS has evolved considerably and become much more oriented to academia than industry.

Given the availability of the IBM 650 at the Watson Lab, the very simple reaction $H + H_2 \rightarrow H_2 + H$, which involves an exchange of hydrogen atom with a hydrogen molecule, could now be studied by theory at a relatively fundamental level. Moreover, early measurements made by **Farkas and Farkas [1935]** of the rate of reaction over a wide temperature range provided data for comparison with calculations.

A second reason for focusing on chemical kinetics was that crossed molecular beam studies were beginning to provide much more detailed information about these reactions than had been available from gas phase or solution measurements. The pioneering experiments of Taylor and Datz opened up this new field in 1955 [**Taylor and Datz, 1955**], although it was not until 2000 that Datz received the prestigious Enrico Fermi Award in recognition of this work. Many groups extended the original crossed molecular beam experiments and showed that it was possible to study individual collisions and determine whether they were reactive or not. This means that calculated reaction cross sections, rather than overall rate constants, could be compared directly with experimental data.

My interest in reaction kinetics was further stimulated by the crossed molecular beam studies of Dudley Herschbach, mainly with his coworker, Yuan Lee. From their beginning in 1961, in what they called the "alkali age," to their extension to a wide range of different systems and different properties [**Herschbach, 2000**], they served to further stimulate my interest in studying chemical reactions by trajectory calculations.

To do a theoretical treatment of this, or any other reaction (including the protein-folding reaction), a knowledge of the potential energy of the system as a function of the atomic coordinates is required, i.e., it is necessary to know the potential energy surface or energy landscape, as it is now often called. Isaiah Shavitt, working with me as a postdoctoral fellow at the Watson Lab on quantum mechanical calculations, had developed new methods for evaluating multicenter two-electron integrals [**Shavitt and Karplus, 1962**], and used the $H + H_2$ potential surface as his first application [**Shavitt et al., 1968**]. Even though this reaction involved only three electrons and three nuclei, the theoretical surface was expected to be useful only for determining the general features of the reaction, while a more accurate surface was required for calculating reaction attributes for comparison with experiment. (Five years later **Liu [1973]** calculated an accurate surface for the H + H_2 exchange reaction.) Thus, we had to resort to a semiempirical model, whose form was given by a quantum mechanical description with parameters determined from experiment. Already in 1936, J. Hirshfelder and B. Topley, two students of H. Eyring, had attempted a trajectory calculation of the H + H_2 reaction with the three atoms restricted to move on a line for simplicity [**Hirschfelder, Eyring, and Topley, 1936**].

In a trajectory calculation one determines the forces, F, on the atoms—here the three hydrogen atoms—from the derivatives of the potential energy surface with respect to the atomic positions, and then integrates Newton's equation, $F = ma$, to obtain the positions of the atoms as a function of time. Hirschfelder and Topley used a three-body potential based on the Heitler–London method. They calculated a few steps along the trajectory but were not able to finish the calculation, so we do not know (and never will since we do not have the initial velocities) whether the trajectory was reactive. The potential had a well in the region where all three atoms were close to each other ("Lake Eyring" as it was called), which was expected to give a three-body complex under the collision conditions appropriate for the reaction. *Ab initio* quantum mechanical calculations, such as that of Shavitt, indicated that this was incorrect, i.e., there was a simple activation barrier. Thus, to obtain a meaningful description of the H + H_2 reaction, a more realistic potential function had to be introduced. Moreover, the potential energy surface had to be accurate in three-dimensional space, so that it would not be necessary to restrict the hydrogen atoms to move on a line.

Richard Porter, a graduate student with F. T. Wall, had used a surface without "Lake Eyring" to improve the collinear collision calculations [**Wall and Porter, 1963; Truhlar and Wyatt, 1976**]. Much impressed by Porter, I invited him to join my group at Columbia as a postdoctoral fellow. At Columbia, we rapidly developed a semiempirical extension of the original Heitler–London surface for the H + H_2 reaction, based on the method of diatomics in molecules and calibrated the surface with *ab initio* quantum calculations and experimental data for the reaction [**Porter and Karplus, 1964**]. This surface, which is known as the Karplus–Porter (KP) surface, has an accuracy and simplicity that led to its continued use over the years in many reaction rate calculations by a variety of methods. It provides another example of the utility of a simple model, which tends to be lost in high level *ab initio* calculations.

Once the surface was developed, we undertook the first full three-dimensional trajectory calculation for the H + H_2 exchange reaction. Dick Porter was a fundamental contributor to the research, which also involved R. D. Sharma, another postdoctoral fellow. With the availability of large amounts (by the standards of the day) of computer time on the IBM 650 at the Watson Lab, we were able to calculate enough trajectories to obtain statistically meaningful results [**Karplus and Porter, 1970**]. Determination of the reaction cross section required the calculation of a series of trajectories with an appropriately chosen range of initial conditions, such as relative velocities and impact parameters. The trajectories start with the reactants far apart (so far that the interaction between the hydrogen atom and hydrogen molecule are negligible), let them collide in the presence of the interaction potential, and then follow the atoms until the products are again far apart. By looking at which atoms are close together, one can determine whether a reaction has taken place. As can be seen from Figure 10.1, the reaction (i.e., the time during which the three atoms are interacting) takes only a few femtoseconds. This illustrates a fundamental point, namely that many simple reactions have a small rate constant, not because of the elementary reaction rate, which is fast when it occurs, but because of the large activation energy, which makes most thermal collisions nonreactive. Within the approximation that classical mechanics is accurate for describing the atomic motions involved in the H + H_2 reaction and that the semiempirical KP surface is valid, a set of trajectories makes it possible to determine any and all reaction attributes, e.g., the reaction cross section as a function of the collision energy. The ultimate level of detail that can be achieved is an inherent attribute of this type of approach, which I was to exploit fifteen years later in studies of the dynamics of macromolecules. The results for the reaction cross section as a function of internal energy and the rate constant as a function of temperature provided insights concerning the fundamental nature of chemical reactions that are as valid today as they were more than fifty years ago when the calculations were performed.

As in many of my papers in which a new method was developed [**Karplus and Porter, 1970**], I tried to present the detail necessary for the reader to reproduce and use what had been done. (We had to work through it all, so why not save others the effort?) The requirement that the work be reproducible is often cited as a standard for publishing papers, but, in practice, few papers are written in this way. I was pleased to learn that our paper was cited by George Schatz [**Schatz, 2000**]

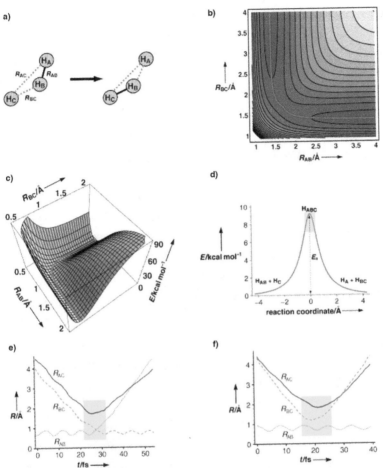

Figure 10.1. *The exchange reaction between a hydrogen atom and a hydrogen molecule. (a) Schematic representation of the reaction with definitions for the distances R_{AB}, R_{AC}, and R_{BC}. (b) Contour plot of the potential energy surface for a linear collision as a function of the distances R_{AB} and R_{BC} with $R_{AC} = R_{AB} + R_{BC}$; the minimum-energy path is shown in red. (c) Same as panel (b), but in a three-dimensional representation. (d) Energy along the reaction coordinate corresponding to the minimum-energy path in panels (b) and (c); the transition state is indicated in yellow. (e) A typical trajectory for the reactive, three-dimensional collision; the distances R_{AB}, R_{AC}, and R_{BC} are represented as a function of time. (f) A typical trajectory for the nonreactive collision, as shown in panel (e). In both panels (e) and (f) the interactions between the three atoms are limited to a very short time period (yellow background); this mirrors the narrow potential energy barrier in panel (d) [Reproduced with permission from Dobson, Sali, and Karplus, 1998]*

in 2000 as one of the key 20th-century papers in theoretical chemistry. Schatz pointed out that what we had called the quasiclassical trajectory method was still widely used. In it, classical trajectories are calculated with quantized initial conditions. This is very important for the H_2 molecule because of its large zero-point vibrational energy, some of which is available in the transition state to cross the activation barrier. It has always seemed fitting to me that Aron Kupperman with his student George Schatz developed quantum scattering methods so that in 1976, essentially ten years after the classical trajectory calculations, they were able to do an accurate quantum calculation; Robert Wyatt with Alan Elkowitz at the University of Texas, Austin, did a corresponding calculation by a somewhat different approach [**Wyatt and Elkowitz, 1975**]. As is pointed out in the 1976 Kupperman–Schatz paper, the classical calculations "agree with the quasiclassical trajectory results of Karplus, Porter, and Scharma (KPS) [**Karplus, Porter, and Sharma, 1965**] within the accuracy of the quantum calculation." This was in some sense a matter of luck, because the $H + H_2$ reaction is essentially adiabatic, so that the correct fraction of the large H_2 zero point energy is available to cross the barrier in the transition state. Moreover, Schatz states [**Schatz, 2000**], "The KPS paper stimulated research in several new directions and ultimately spawned new fields." A very nice comment to have thirty-five years after the work was done.[2]

After the $H + H_2$ reaction calculation, Dick Porter and I also collaborated on the textbook *Atoms and Molecules* [**Karplus and Porter, 1970**], which developed quantum chemistry at the introductory level for students of physical chemistry. It was based on a lecture course I had given several times at Columbia and then at Harvard University after I left Columbia. Teaching the course was very good preparation for writing the book, because it required me to think carefully about the subject so I could present it clearly enough for students to understand. Dick and I began working on the book only after I had moved to Harvard. He joined me at Harvard for one semester to teach the physical chemistry course, which gave me extra time to devote to writing. We assumed that writing a book based on the lecture notes would be simple to do, but in fact it was an enormous task. The book was finished only because Dick and I were together for a summer on Martha's Vineyard at a "writing camp" for scientists, sponsored by an imaginative young publisher, Bill Benjamin. We were housed in a large mansion on the

---

[2]While I was on the faculty at Columbia, I received a somewhat surprising request from a lawyer working for Professor Pauling. Pauling was suing *The New York Daily News*, which had referred to him as a Communist, for defamation of character. The lawyer asked me to serve as a character witness. Of course, I agreed to do so. On the stand, I was nervous but did manage to express my strong support for Pauling and to make clear that, although he was outspoken against the testing of atomic bombs in the atmosphere, he was not a Communist. Several well-known scientists also testified on behalf of Pauling. Pauling lost the case, the judge deciding that since Pauling had such people testifying on his behalf, it meant that he had not suffered any injury from *The Daily News* article.

coast, worked together in the morning, and were free to go swimming or sailing, or just wander around in the afternoon. The text of the book was essentially finished that summer. Subsequently, while spending a semester at the Weizmann Institute in Rehovot, Israel, I used part of the time to correct the proofs. In my case, this inevitably led to significant rewriting. *Atoms and Molecules* has been a success, in part, because it never used the phrase "It can be shown" (Figure 10.2).

ATOMS &
MOLECULES:
An Introduction
For Students of Physical
Chemistry

Martin Karplus and Richard N. Porter

Figure 10.2. Cover of Atoms and Molecules

It is still in use, particularly as a source of material by teachers of physical chemistry.[3]

Unlike quantum dynamics calculations, the quasiclassical trajectory method was easily extended to more complex systems. One study that I remember as particularly interesting was done by Martin Godfrey. It concerned the $K + CH_3I \rightarrow KI + CH_3$ reaction, for collisions involving an orientated CH_3I molecule. We were stimulated to do this calculation by an ingenious experiment performed by Richard Bernstein, who was an outstanding contributor to the field of crossed molecular chemistry. He was able to orient the CH_3I reactant relative to the incoming K^+ ion so that one could study the effect of the orientation on the reaction and obtain additional information for comparison. The two papers describing the work were published together in the *Journal of the American Chemical Society* [**Beuhler, Bernstein, and Kramer, 1966; Karplus and Godfrey, 1966**].

---

[3]One of my disappointments is that I started two other books, spent much time on both them, but, unlike *Atoms and Molecules*, did not finish either one. Of these, one early in my career, concerned the revision of the *Introduction to Quantum Mechanics* by Pauling and Wilson. Shortly after completing my PhD, Pauling invited me to update this classic text, which was originally published in 1935. I think Pauling did so to help me to become better known and assumed I would do a mild "makeover," which would take little time. However, like many of my publications, I felt I had to do the best job possible and began to do an extensive revision. My draft went through about half of the book by 1974, when it became clear to me that it was unlikely to be completed. I felt that Pauling felt that in a certain sense, I was destroying what was a classic text, even if it was somewhat out-of-date. Pauling wrote me in April 1974, that if we were to go ahead, the "revised" version should appear as a new book and the original should continue to be available (see also page 168).

Chapter 11

The FBI and I

———————— ✥ ————————

My interactions with the FBI appear to have begun when I was a graduate student at Caltech. I had participated in 1952 at a meeting concerned with organizing a protest movement against the death penalty sentences of Julian and Ethel Rosenberg.[1] At the entrance to the meeting hall, there was a desk at which we were asked to write down our contact information and make a contribution. Apparently, an FBI agent was there to make a list of the participants. I first discovered this when, in the mid-1970s, my brother Bob was refused a security clearance to do research at the Livermore Laboratory. Because his participation was deemed important to the laboratory, the administrators appealed and found out that the reason for the refusal was my documented presence at a Rosenberg protest meeting two decades previously at the First Unitarian Church in Los Angeles led by the liberal minister, Stephen Fritchman. Once this unimportant item in his FBI file was found, Bob rapidly received the necessary clearance.

In contrast to my brother's experience, the one time I needed a clearance it was not granted. I was invited to participate in a disarmament study sponsored in Woods Hole by the National Academy of Sciences in the early 1970s. Every participant had to have a security clearance so that when the report was issued, no one could claim that the participants did not have access to the information concerned with secret government negotiations. I arrived in Woods Hole expecting to receive clearance and attended the opening meeting, which was public. After that, I was not allowed in the sessions while awaiting my clearance, which never arrived. During the month-long meeting, I relaxed with my family and worked on my own research in a pleasant nearby hotel with a swimming pool, courtesy of the sponsors. This also permitted me to renew my ties with the Marine Biological Laboratory.[2]

———————— ✥ ————————

[1]Ethel and Julius Rosenberg, both American citizens, were convicted of passing atomic bomb secrets to Russia on April 5, 1951, and were sentenced to be executed. Many Americans believed that the death sentence was, at least in part, a response to the hysterical fear of Communism generated by Senator Joseph McCarthy. During the two-year period between the sentencing and their execution on July 19, 1953—the Rosenbergs were the first US citizens to be convicted for espionage and executed during peacetime—there was a worldwide effort to have their sentences reduced.

[2]The disturbing element in my FBI record was the one that had raised problems for my brother, namely that I had attended the Rosenberg protest meeting. It amazes me that the US government wastes so much time and money recording the attendance

Another experience with the FBI occurred during the years 1955–1956, while I was on the faculty of the University of Illinois. Shortly after I moved to Urbana, I began to receive visits from FBI agents. There were always two of them: one was from the local (Chicago) office and the other was a different person each time. They questioned me about my political views and on several occasions focused on my visit to Yugoslavia while I was a postdoctoral fellow at Oxford (see Chapter 17). On one such visit, the "second" agent suddenly began speaking in a language I did not understand. It turned out to be Serbian and was apparently meant to trick me. What exactly the FBI was looking for, I never discovered. The visits stopped after about six months.

At various times starting in early 2011, I applied to the FBI for my file under the Freedom of Information Act (FOIA, signed into law by President Lyndon Johnson in 1966), but only received replies that were clearly intended to discourage further inquiries. I was notified that there was a file on me, but repeated requests to receive the file elicited no response. During my visit to the White House shortly after the Nobel Prize had been announced in October 2013 (see Chapter 21), I mentioned to John Holdren, the Science Advisor to President Obama, that I had had no response to my FOIA request. Following efforts on my behalf by Holdren and his office, I received the 213-page file in March 2014 (Figure 11.1).

Until I received the FBI file, I did not realize that I was CASE Number E1031981 with the classification "subversive" and that the file recorded in detail the various interviews at the University of Illinois going back to July 1956 when that report was filed. I was asked to fill in form DD398, which is a Statement of Personal History, at a number of the interviews at Illinois but refused to do so according to the record. Apparently I said that it was not germane and that listing character references would inconvenience people unnecessarily. Also, I evidently stated that I was against Communism as it was practiced in Russia, but did not agree with the actions of Senator McCarthy and the House on Un-American Activities Committee in the United States. Rereading the report of the various agents who came to see me in Urbana, I felt sorry for them. It is clear that they were just trying to do their assignment, while I showed my annoyance with them.

---

at meetings, which one should have the right to attend without fear of future harassment or discrimination. As to the Woods Hole meeting, its report was declassified and made public soon afterward.

NATIONAL ARCHIVES *and* RECORDS ADMINISTRATION
8601 ADELPHI ROAD COLLEGE PARK, MD 20740-6001
www.archives.gov
NATIONAL ARCHIVES

March 12, 2014

Martin Karplus
Harvard University
Department of Chemistry & Chemical Biology
12 Oxford Street
Cambridge MA 02138

NARA Case Number: **RD 36499**

Dear Mr. Karplus:

This is in further response to your Freedom of Information Act (FOIA) request for access to records in the custody of the National Archives and Records Administration (NARA). Our office received your request for Federal Bureau of Investigation case file 116-HQ-450978 on September 21, 2011. This file, totaling 213 pages, is part of Record Group 65, Records of the Federal Bureau of Investigation; Entry Ud-09D 5, Classification 116 Headquarters Case Files; box 50.

We have completed our review. 206 pages are released in full. 7 pages are released with portions withheld pursuant to exemption (b)(7)(D) of the Freedom of Information Act (5 U.S.C. 552). You or your representative may view this material in the research room at the National Archives at College Park; please contact us at least 72 hours in advance of your visit. If you wish to order copies of this material by check or money order made payable to the "National Archives Trust Fund," forward your payment with a completed copy of the enclosed quotation for reproduction services form (Form 72) to: National Archives Trust Fund, Cashier (BCT), Form 72 Order, 8601 Adelphi Road, College Park, Maryland 20740-6001. We only accept payments made in U.S. dollars and drawn from U.S. banks. If you prefer to pay using American Express, MasterCard, Visa, or Discover credit card, fax the completed form with your credit card information to 202-357-5906 or call the Accounts Receivable Specialist at 301-837-3720 and provide your credit card information. Retain one copy of the enclosed Form 72 for your records.

If you are not satisfied with our response to your request, you may administratively appeal by writing to the Deputy Archivist of the United States (ND), National Archives and Records Administration, 8601 Adelphi Road, College Park, Maryland 20740-6001. The Deputy Archivist must receive you appeal within 35 days of the date of this letter. Clearly mark the appeal letter and envelope "FOIA Appeal" and cite your case number in your appeal letter.

This concludes the processing of your request. If you have any question, please contact James Mathis at james.mathis@nara.gov or (301) 837-0288.

Sincerely,

MARTHA WAGNER MURPHY
Chief
Special Access/FOIA Staff

Figure 11.1. Letter confirming that I would receive my FBI file

The file consists primarily of interviews with many people who have known me, some of whom I had forgotten, as well as information on some organizations with which I had been involved. For example, the file states that "a special investigative technique" had revealed that I had received mail in Newton in the mid-1950s from the Central Committee for Conscientious Objectors, which as the FBI stated "unwittingly allowed itself to be Communist used." It also refers to the FBI investigation regarding the accusation of my father's supposed Nazi sympathies mentioned earlier (see Chapter 4). In some ways, the file could be used as part of my autobiography with references to favorable statements about me from friends, colleagues, mentors, and even landladies.

At least 100 people were contacted by more than twenty different agents. The file lists each person interviewed as a "source." Typical are some sentences from the interview with Verner Schomaker, who was a Caltech professor when I was a graduate student (see Chapter 7): "From 1952, source met SUBJECT [Capitals from original] on the average of once a month at dances and parties organized by him and his roommates." "Source added that Karplus is one of the best students in the field of physical chemistry that has graduated from Caltech in the past ten years." "He added Karplus is completely honest, reliable, of high integrity, and he has no reason to doubt SUBJECT'S loyalty to the United States." The FBI report finished with a list of people who knew the SUBJECT, each of whom were interviewed in turn. There was also a "NEIGHBORHOOD CHECK," which involved individuals living near my parents' house at 259 Otis Street in Newton, Massachusetts. Their comments were generally in contrast to those of my scientific colleagues. "SUBJECT's parents have never participated in neighborhood social activities, possibly because they are of Jewish extraction." "One rumor about the family, which might have a bearing on SUBJECT's father's trouble with the immigration authorities was that he was allegedly a refugee from Austria but entered this country with $9,000 and was able, almost immediately, to purchase a nice house in an expensive neighborhood."

Of particular interest to my interviewers was the fact that I had received my PhD with Linus Pauling. Twelve pages in my FBI file were devoted to him. The file included a list of about 200 organizations designated by the Attorney General (Exec Order 10450, December 1950) membership in which was a ground for suspicion. Examples that concerned Pauling were his membership in the "Emergency Committee of Atomic Scientists" and his chairmanship of the "West Committee to Welcome the (Red) Dean of Canterbury." Also, my membership in the American Civil Liberties Union (ACLU) was a concern, as described over three pages. The file states "From its inception in 1920 until 1940 the ACLU was sympathetic to the government of the Soviet Union and apparently blind to the objectives of the CP in the United States." The chief criticism of the ACLU was that it devoted most of its resources to the defense of Communists and leftist causes. However, it was not included in the list of subversive organizations mentioned earlier.

The investigation ended in 1962 with the file not listing any conclusion about me. The only statement was "The file is so long that only three copies will be submitted." On the first page of the file it states in capital letters: "DO NOT DESTROY HISTORICAL VALUE NATIONAL ARCHIVES." This leads me to wonder whether anyone else will ever be interested.

Chapter 12

Return to Harvard University and Biology

———— ⚭ ————

In 1965, it was time to move again. Columbia and New York City were stimulating places to live and work, but I felt that new colleagues in a different environment would help to keep my research productive. I had incorporated this idea into a "plan": I would change schools every five years and when I changed schools I would also change my primary area of research. It was more exciting for me to work on something new where I had much to learn. Following this plan helped me to stay mentally young and have innovative ideas, in part because I was not encumbered by knowing all the literature in the field. Since I generally decided to start some research in areas where my new colleagues were doing experiments, the change in fields was also usually a plus for the institution.

When it became known that I was considering leaving Columbia University, numerous schools invited me to join their faculty. With considerable difficulty, I narrowed the choice down to U. C. Berkeley and Harvard and decided to visit each place for a semester during my 1965–1966 sabbatical year. Before taking this leave, I made sure that Columbia would not require me to come back for a year or more after the sabbatical; such a requirement was often a condition of being allowed to take a sabbatical. I enjoyed my stay in both places; discussions with colleagues were stimulating, and it was very hard to decide between the two schools. I particularly enjoyed my interactions at U. C. Berkeley with Bob Harris, a gifted theoretician who ten years before had been my first research student while he was an undergraduate at the University of Illinois. We spent many hours together talking science and politics. This was the era of the Vietnam antiwar movement, and I was introduced to police brutality during some of the marches in Berkeley, a center of the movement, and particularly in neighboring Oakland, where there was strong support for the Vietnam War, in contrast to Berkeley.[1]

———— ⚭ ————

[1]One of the marches in which I participated went through Oakland. I took photographs of the police beating student marchers with their batons. Because I used a technique, described in Chapter 17, the police were not aware that I was photographing them and so left me alone. I mailed the slides to the *Oakland Tribune*, but they were not published. Unfortunately, I had not made copies, though I had kept some slides of the marchers (Figure 12.1).

Figure 12.1. Protesters against the Vietnam War

A variety of factors, conscious and unconscious, led me to choose Harvard. The fact that I had had such a positive experience as a Harvard undergraduate influenced me. Ironically, I felt that the Berkeley environment and its weather were just too nice and that the distractions from work were too great, as evident from the activities of some of the Berkeley chemistry faculty at that time. One faculty member seemed to spend more time on his ranch than in research.

I joined the faculty at Harvard as professor in the fall of 1966 and received the title of Theodore William Richards Professor in 1979. Although such chairs then meant little at Harvard, other than that the funds for one's salary come out of a special endowment, I was very pleased to receive this particular title. First, the previous holder of the chair had been E. Bright Wilson. He was a highly respected member of the department for his science as well as for his humanity in dealing with students, and particularly for his high standards of intellectual honesty. Second, this chair was the only one in chemistry named after a scientist instead of the donor of the funds.[2]

---

[2]T. W. Richards was the first American scientist to receive a Nobel Prize in Chemistry. It was awarded in 1914 for the exact measurement of atomic weights of a large number of elements. The T. W. Richards chair was endowed by Thomas Lamont in 1925 to honor his brother's friend, T. W. Richards. The sum involved was $100,000. Today, endowment of a chair requires about five million dollars [W. Bentinck-Smith and E. Stouffer, *Harvard University History of Named Chairs* (Cambridge, Massachusetts, 1991)].

While I was at the University of Illinois, the possibility of being a junior faculty member at Harvard arose, but I decided to go to Columbia, primarily because I felt that the stress of wondering whether I would be promoted to tenure would have been too much for me. At that time, many of the junior faculty were not given tenure, or as one says, "saw the writing on the wall" and left even before a tenure decision was made. My brother Bob was, in fact, an assistant professor in the Department of Physics at Harvard from 1950 to 1954, when he moved to U. C. Berkeley as a tenured associate professor. The argument for not promoting people to tenure was that the department wanted to find the best person in the field and the fact that one was already at Harvard played only a small role. One criterion for promotion to tenure was that one had published a paper that was likely to remain important for ten or more years. In that sense, my work on NMR, which led to the "Karplus equation" and the $H + H_2$ reaction quasiclassical trajectory calculations both satisfied the criterion. They presumably played a role when I was finally invited back to Harvard as a professor in 1965.

At Harvard I continued to do research in some of the areas that I had developed at Illinois and Columbia, including the study of hyperfine interactions in ESR [**Purins and Karplus, 1969**] and the use of quasiclassical trajectory methods for the study of reactions [**Godfrey and Karplus, 1968**]. With the results from the trajectory calculations, Keiji Morokuma, B. C. Eu, and I undertook a study of the relation between reaction cross sections and transition state theory as a test of this widely used model in chemistry [**Morokuma, Eu, and Karplus, 1969; Morokuma and Karplus, 1971**]. The application of many-body theory to the electronic structure of atoms and molecules [**Caves and Karplus, 1969; Freeman and Karplus, 1976**], as an extension of methods developed in physics, was also of interest to me, in part to better understand what was involved in this development. I find it very helpful when I see a new idea or method to apply it to some problem, even if the specific research that I do is not always highly significant.

After I had been at Harvard for only a short time, I realized that if I was ever to return to my long-standing interest in biology I had to make a break with what had been thus far a successful and very busy research program in theoretical chemistry. Also, I felt that I understood magnetic resonance and elementary chemical reactions and that, as a result, my excitement about possibly learning something new in those areas was no longer there. The initial qualitative insights obtained from relatively simple approaches to a new problem are often the most rewarding. This is not to imply that the field of gas-phase chemical reactions has not continued to flourish. It is still active, with ever-finer details concerning reactive collisions being elucidated. I very much enjoyed, for example, attending a meeting on the dynamics of elementary reactions at the Fritz Haber Institute in Berlin in 1982, where I learned about the exciting research going on. My lecture, however, was on dynamics of protein [**Karplus, 1982**]. The meeting also brought home to me that other people with skills different from mine were better able to contribute to the advanced technologies now required in this area.

I decided to take a six-month leave in the fall of 1969 and chose the Weizmann Institute, in part because it had an excellent library. I was aware of Shneior Lifson's work on polymer theory and his reputation of being an open-minded scientist, as well as a marvelous storyteller. I wrote Lifson asking whether I could come for a semester, and he invited me to join his group. The sabbatical gave me the possibility, in addition to going over proofs for *Atoms and Molecules* (see Chapter 10), to read and explore a number of areas in which I could hope to do constructive research by applying my expertise in theoretical chemistry to biology. Discussions with Lifson and the many visitors to his group helped me in these explorations.

A key, although accidental, element in my choice of the first biological problem was the publication of *Structural Chemistry and Molecular Biology*, a compendium of papers in a volume dedicated to Linus Pauling for his 65th birthday. I had contributed an article entitled "Structural Implications of Reaction Kinetics," which reviewed some of the work I have already described in the context of Pauling's view that a knowledge of structure was the basis for understanding reactions [**Karplus, 1968**]. However, it is not this article that leads me to mention the volume, but rather an article by Ruth Hubbard and George Wald entitled "Pauling and Carotenoid Stereochemistry." They reviewed Pauling's contribution to the understanding of polyenes with emphasis on the visual chromophore, retinal. The article contained a paragraph, which I reproduce here because it describes an element of Pauling's approach to science that greatly influenced me:

> One of the admirable things about Linus Pauling's thinking is that he pursues it always to the level of numbers. As a result, there is usually no doubt of exactly what he means. Sometimes his initial thought is tentative because the data are not yet adequate, and then it may require some later elaboration or revision. But it is frequently he who refines the first formulation.

The article on retinal made clear to me that the theory of the electronic absorption of retinal and its geometric changes on excitation, which play an essential role in vision, had not advanced significantly since the discussions I had with Hubbard and Wald during my undergraduate days. I realized, in part from my time with Coulson (I had learned something from him in Oxford after all!), that polyenes, such as retinal, were ideal systems for study by the available semiempirical approaches. If any biologically interesting molecule in which quantum effects are important could be treated adequately at that time, retinal was it. Barry Honig, who had received his PhD in theoretical chemistry working with Joshua Jortner, joined my research group at that time. He was the perfect candidate to work on the retinal problem. It was known that 11-cis retinal, the chromophore which absorbed the quantum of light, is not planar. It is twisted about the C_{12}-C_{13} single bond, and this was thought to play a role in the photoisomerization reaction (the C_{11}-C_{12} double bond changes from *cis* to *trans*) that gives rise to the visual signal. There was no three-dimensional structure of retinal, and, in particular, it was not known whether the twist led to an 12-s-*cis* or 12-s-*trans* configuration (Figure 12.2).

Barry Honig did a calculation with a Hückel one-electron Hamiltonian for the π-electron system and a pairwise nonbonded energy function for the σ-bond framework of the molecule [**Honig and Karplus, 1971**]. The theory predicted that the structure was 12-s-*cis*. Perhaps even more important than the theoretical result *per se*, is the fact that it was the first paper in which part of the system

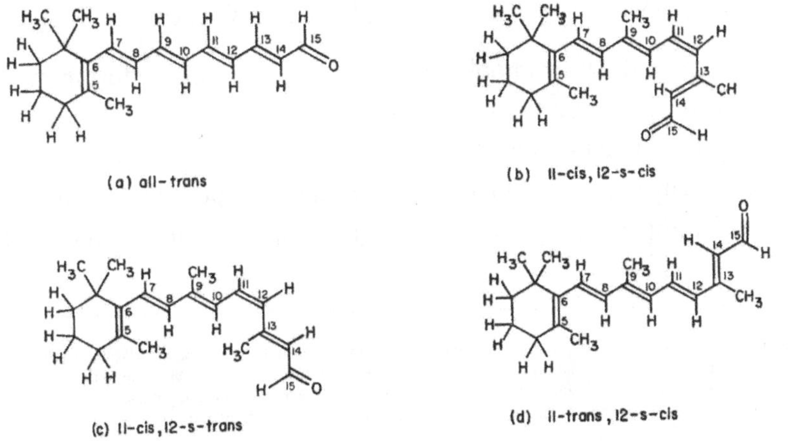

(a) all-trans

(b) ll-cis, 12-s-cis

(c) ll-cis, 12-s-trans

(d) ll-trans, 12-s-cis

Figure 12.2. *The different forms of retinal considered in the calculations* (here the σ-bond framework) was treated by a classical mechanical model and the π-electrons, with which the interaction with light takes place, were treated by quantum mechanics, albeit by a very simplified model. This type of treatment was chosen by the Nobel Committee as the basis of the Nobel Prize awarded in 2013 to me, with Michael Levitt and Arieh Warshel, as stated in the Nobel citation: "for the development of multiscale models for complex chemical systems."

Honig and I felt that our result for the structure of 11-*cis* retinal with its implication for visual excitation was appropriate for publication in *Nature*. We submitted the paper, which received excellent reviews, but it was sent back with a rejection letter stating that because no experimental evidence supported our results, it could not be verified that the conclusions were correct. This was my first experience with *Nature* and with the difficulty of publishing theoretical results related to biology, particularly in "high impact" journals, such as *Nature*, and *Cell*. If theory agrees with experiment, it is not particularly interesting because the result is already known. Whereas if one is making a prediction, it is difficult to publish because it could be incorrect. Various aspects of this publication problem are reviewed in a paper I published with Richard Lavery in an issue of the *Israel Journal of Chemistry* commemorating the 2013 Nobel Prize [**Karplus and Lavery, 2014**].[3] Upset by this rejection letter, I called John Maddox, the editor of *Nature*, and explained the situation to him. Apparently, I was successful and the paper was accepted. A subsequent crystal structure verified our prediction [**Gilardi et al., 1971**]. In a review of studies of the visual chromophore [**Honig, Warshel, and Karplus, 1975**], I noted that, "Theoretical chemists tend to use the word 'prediction' rather loosely to refer to any calculation that agrees with experiment, even when the latter was done before the former."

The study of the retinal chromophore gave rise to a sustained effort in my group concerned with the properties of retinal and other polyenes. It was fostered, in addition, by Bryan Kohler and Bryan Sykes, two assistant professors who had joined the Chemistry Department and were doing experiments that provided challenges for theory. They were part of a group of young faculty that also

---

[3]In at least one of the most prestigious journals, *Cell*, I had the good fortune that the editor, Emilie Marcus, wanted to publish an article I had submitted. At first she did not see how *Cell* could publish a theoretical article, since up to then it had only published experimental papers. She finally had the idea of introducing a "Theory" section and published it there [Gao, Yang, and Karplus, 2005]. The Theory section is now quite robust and articles appear regularly in it.

included William Reinhardt and Roy Gordon, plus Bill Miller, a junior fellow, which made the department particularly stimulating at that time. Our offices were located along a narrow corridor on the ground floor of Converse Laboratory, with my office at one end and Roy Gordon's at the other. My daily strolls down this corridor provided many occasions for scientific discussion. A collaboration with Sykes led to one of the earliest uses of vicinal spin–spin coupling constants and the nuclear Overhauser effect in NMR to determine the conformations of a biomolecule (retinal in this case) [**Honig et al., 1971**]. The technique, in a much more elaborate implementation, is now the basis of the use of NMR for protein structure determinations, a method pioneered by Wagner and Wüthrich. Kohler and his student, Bruce Hudson, were doing high-resolution spectral studies of polyenes, such as hexatriene. They had observed a very weak absorption lower in energy than the strongly absorbing transition, which is the analog of the one involved in retinal isomerization. It was suggested that there exists a "forbidden" transition, which was not predicted by the simple (single-excitation) models of polyene spectra, such as the one used by Honig in our study of retinal. Klaus Schulten, then a graduate student, who work jointly with Roy Gordon and me, introduced double excitations into the Parisar-Parr-Pople (PPP) approximation for π-electron systems and found the low-lying (forbidden) state in hexatriene and octatetraene [**Schulten and Karplus, 1972**]. A number of related studies followed. Arieh Warshel had joined my group at Harvard after we met at the Weizmann Institute, where he had been a graduate student with Lifson. He extended the polyene model by introducing a quantum mechanical Hamiltonian that refined the PPP method for the π-electrons and by treating the σ-bonded framework by a more elaborate molecular mechanics approach fitted to a large set of experimental data [**Warshel and Karplus, 1974**]. The method, like the simpler model used by Honig, was an early version of the quantum mechanical/molecular mechanical (QM/MM) approach that came to be widely employed for studying enzymatic reactions [**Field, Bash, and Karplus, 1990**]. At the time, we used the method to calculate the vibronic spectra of retinal and related molecules [**Warshel and Karplus, 1974**]. Subsequently, a collaboration was initiated with Veronica Vaida, a member of the chemistry faculty, and her graduate student Russ Hemley. He extended the approach we had developed to excited states of molecules like styrene [**Hemley et al., 1985**], which Vaida and her students were studying experimentally.

In the 1970s, I moved my office and those of my research group from Converse to Mallincrodt, where the large lecture hall had been renovated into a three-story integrated space to house the physical chemistry faculty and the theoretical students. The renovated area was known as the "New Prince House"; Prince House was an old Cambridge house near the Chemistry Department where theoretical students had offices for a number of years (Figure 12.3).

It promoted interaction among all occupants—senior and junior faculty, whose offices were located on the upper and mid-tier, and the theoretical postdocs and graduate students who had offices in the lower depths of the tri-level complex. Its lounge area, equipped with an espresso coffee machine, provided a congenial environment ideal for generating discussions. Among the many interactions over the twenty-year period during which the complex existed, none proved more fruitful than those with Chris Dobson, who was a junior faculty member in the department from 1978 to 1980 before returning to Oxford. Our collaborations continued after he returned to Oxford and then moved to Cambridge.

Figure 12.3. Prince House

A biological question that appeared ready for a more fundamental investigation was hemoglobin cooperativity, the model system for allosteric control in biology. Although the phenomenological model of Monod, Wyman, and Changeux [**Monod, Wyman, and Changeux, 1965**] had provided many insights, it did not attempt to make contact with the detailed structure of the molecule. I had already begun working on hemoglobin with Robert Shulman, then at Bell Labs, who had measured the paramagnetic NMR shifts of the heme protons and we had developed an interpretation of the results based on the electronic structure of the heme [**Shulman, Glarum, and Karplus, 1971**]. In 1971 Max Perutz had just determined the X-ray structure of deoxy hemoglobin, which complemented his earlier results for oxy (actually met) hemoglobin [**Perutz, 1971**]. By comparing the two structures, he was able to propose a qualitative molecular mechanism for the cooperativity. Alex Rich, then a professor at the Massachusetts Institute of Technology, invited Perutz to present two lectures describing the X-ray data and his hemoglobin mechanism. After the second lecture, Alex suggested that I come to his office to meet Perutz. He was sitting on a couch in Alex's office eating a banana, as was his custom. (He had stomach problems that severely restricted his diet). I asked Perutz whether he had tried to formulate a quantitative thermodynamic mechanism based on his structural analysis. He said no and seemed very enthusiastic, although I was not sure whether he had understood what I meant. Having been taught by Pauling that until one expressed an idea in quantitative terms, it was not possible to test one's results, I went away from our meeting thinking about the best way to proceed. Attila Szabo had recently joined my group as a graduate student, and an investigation of the hemoglobin mechanism seemed like an ideal problem for his theoretical skills. The basic idea proposed by Perutz was that the hemoglobin molecule has two quaternary structures, R and T, in agreement with the ideas of Monod, Wyman, and Changeux; that there are two tertiary structures, liganded and unliganded for each of the subunits; and that the coupling between the two is introduced by certain salt bridges whose existence depended on both the tertiary and quaternary structures of the molecule. Moreover, some of the salt bridges depended on pH, which gave rise to the Bohr effect on the oxygen affinity of the subunits. These ideas were incorporated into the statistical mechanical model Szabo and I developed [**Szabo and Karplus, 1972**]. It was a direct consequence of the formulation that the cooperativity parameter n (i.e., the Hill coefficient) varied with pH. This was in disagreement with the hemoglobin dogma at the time and led a number of the experimentalists in the field to initially disregard our model; careful measurements of the dependence on pH subsequently confirmed our conclusions.

When we began working on the model, I discussed our approach with John Edsall and Guido Guidotti, both biology professors at Harvard. Edsall was well known for his deep understanding of protein thermodynamics and Guidotti was an expert on hemoglobin. There were a number of parameters in the model and we had chosen their values by use of physical arguments. Because the values of the parameters were estimated, the results from the model were only in approximate agreement with experiment. Guidotti warned me that such results would not be accepted by the hemoglobin community, in particular, and biologists, in general. Consequently, we inverted the description of the model. We used experimental data to determine the parameters so that the agreement with experiment was excellent and then justified the values of the parameters with the physical arguments we had developed. During the formulation of our ideas, we often asked Guidotti which of certain experiments were to be trusted, since the nearly overwhelming hemoglobin literature contained sets of data that disagreed with each other.

Chapter 13

Move to Paris

———————— ⊚❧⊚ ————————

The paper describing the hemoglobin work was written in Paris, much of it at Café Les Deux Magots, a left-bank café famous as a rendezvous for writers and philosophers such as Jean-Paul Sartre. I was on sabbatical leave in 1972–1973 and officially at the Université de Paris XI in Orsay, a suburb of Paris, associated with the group of Jeannine Yon-Kahn, a pioneer in experimental studies of protein dynamics. However, I was living (at that time with my wife Susan and our two daughters Reba and Tammy) in the heart of Paris in the 5th arrondissement. Since the trip to Orsay by the Réseau Express Régional (RER) (commuter rail) took about 45 minutes, I spent much of my time at the Institut de Biology Physico-Chimique on rue Paul et Marie Curie in the 5th arrondissement. Having found Paris a wonderful place to live during sabbatical visits, I began to consider the possibility of moving there on a permanent basis. I had been at Harvard for the canonical five years, and the idea of returning to live in Europe was tempting. Given my recollection of the escape from Austria and the Nazi-leaning parties still prevalent there, I had no inclination to return to the country of my birth.[1] France offered many attractive aspects of European life and culture, and I believed that I could do high-level research in theoretical chemistry and its biological applications in Europe, as well at Harvard. After the 1968 revolution, the immense Université de Paris, with more than 300,000 students, had been divided into a dozen campuses. With true French rigor, they were named Paris I, Paris II, and so on, although they now have names, in addition to their numbers.

Orsay (Paris XI), where I had spent my sabbatical, was one of three science campuses, and certainly the best. However, if I was going to move to France, I wanted to live in Paris itself. Consequently, I focused on the two other scientific universities (Paris VI and Paris VII) that were intertwined in the 5th arrondissement on the Jussieu campus, a block of ugly modern buildings. A saving grace was their central location in the area where the Halles aux Vins had been located

———————— ⊸⧉⊷ ————————

[1] The German Chancellor Willy Brandt apologized in the 1970s, for the Nazi atrocities but it was only in the early 1990s that the Austrian government formally recognized that it was not a victim of Germany and accepted its responsibilities for what the Nazis did.

before the Second World War. They had been partly destroyed by bombings in 1944. However, the neighboring streets were still dotted with good inexpensive restaurants dating back to the area's previous existence and now thriving on the faculty and student clientele.

In discussions with European colleagues who had encouraged me to make the move to France, a serious obstacle became clear. I was a tenured professor at Harvard and, needless to say, was willing to move only if I was offered a permanent position in Paris. However, French university professors were civil servants and only French citizens could be civil servants. At the time there was no way of obtaining French citizenship without losing American citizenship. (It is now possible and our son, Mischa, has dual citizenship.) However, many things could be achieved in France by someone with political influence. Jacques-Emile DuBois, a chemistry professor at Paris VII with connections to the Department of Defense of the French government, said he would try to "arrange the situation." I did not know exactly what he meant but hoped that the tenure problem could be solved. On that basis I took a leave of absence from Harvard.

With only this verbal commitment of a permanent position, I moved the majority of my research group, including David Case, Bruce Gelin, and Iwao Ohmine, among others, from Harvard in the fall of 1974. At Paris VII, empty laboratory spaces awaited us. We were provided with some funding by the university and bought the necessary office furniture and computing equipment and went to work. The transition was made much easier with the arrival of Marci Hazard, who had joined the group as secretary in May. (Seven years later we were married.) Many of the logistical problems (e.g., finding where to purchase what we needed) were solved by her, and she played a key role in the cohesion of the group.

As the year went on, DuBois reported on his progress in regularizing my status. Initially, I was appointed Professeur Associé, which is an annual appointment open to non-French citizens. Finally, in January 1975, a decree was published in the official government register (much of the French government, not unlike that of the United States, functions by presidential decrees, without requiring approval by the National Assembly) exempting university professors from the citizenship requirement. This made possible my appointment as a tenured professor. It was only in April that I received formal notification of my university status. However, not everything had been resolved, and the complexity of dealing with the French administration over this and other matters led me to renounce my dream and return to Harvard with my group at the end of the sabbatical year.[2]

During this period I started spending summer vacations with my family in the foothills of the Alps above Annecy and its stunning lake, an area that I had first seen on my postdoctoral trips in the early 1950s. My colleagues at Harvard viewed such absences from Cambridge as improper. However, I found that being away gave me a chance to think, undistracted by everyday pressures. Hikes in the mountains provided the backdrop for my reading and thinking and played an essential role in developing new areas of research.

Faculty members at Harvard are paid annual nine-month salaries for their classroom teaching and are allowed to pay themselves a "summer salary" for the remaining three months using research grants. I find this system annoying, as well as hypocritical, since, in fact, most of our teaching has to do with advising graduate students or postdoctoral fellows. With this in mind, I asked my National Science Foundation (NSF) program director whether the summers in the Haute Savoie were justified under the conditions of my grants. His conclusion was that, given their importance to my research, they were an appropriate, if somewhat unorthodox, summer program.

In 1974, I purchased a plot of land with a magnificent view of the surrounding mountain ranges. It was in Chalmont, a small hamlet in the Manigod valley above Lac d'Annecy. A chalet was built, which was our summer home for more than thirty years (Figures 13.1 a–d).

As soon as we had boarded the plane, I felt that I had left many cares behind. The rural environment, afternoons spent talking with one of our neighbors, who we called the "old farmer" (Figure 13.2a), about the haying still done by hand (Figure 13.2b) and other local events, going swimming at a beach in nearby Talloires (Figures 13.3 a and b), as well as hiking (Figure 13.4), did wonders for all of us.

---

[2]The decree, however, remained valid and subsequently I received a number of thank-you letters from non-French scientists who had for many years been appointed annually as Professeur Associé and suddenly received a permanent position. Jean-Pierre Hanson, who was born in Luxembourg, told me that he is one of the first people to have profited from "my" decree.

(a)

(b)

(c)

(d)

Figure 13.1. Our chalet in Chalmont. (a) Chalet in the summer. (b) Chalet in the winter. (c) View from chalet. (d) Marci, Mischa, and I on the porch of the chalet, 1983

(a)   (b)

Figure 13.2. (a) The "old farmer" with Mischa, 1984. (b) Me haying in Chalmont, 1988

(a)   (b)

Figure 13.3. (a) Talloires, 1984. (b) Talloires beach, 1984

Figure 13.4. On a hike with Reba and Tammy, 2016

Chapter 14

Protein Folding, Hemoglobin, and the CHARMM Program

———— ⚬❦⚬ ————

C y Levinthal, had pointed out, in what came to be known as the Levinthal paradox, that to find the native state by a random search of the astronomically large configuration space of a polypeptide chain would take longer than the age of the earth, while proteins fold experimentally on a timescale of microseconds to seconds or minutes. When Chris Anfinsen visited Rehovot, while I was in the Lifson group in 1969, I became aware that the mechanism of protein folding posed a difficult problem. He described the experiments that had led to the realization that many proteins can refold in solution, independent of the ribosome and other parts of the cellular environment [**Anfinsen, 1973**].[1] What most impressed me was Anfinsen's film showing the folding of a protein with "flickering helices forming and dissolving and coming together to form stable substructures." The film was a cartoon, but it led to my asking him, in the same vein as I had asked Perutz earlier about hemoglobin, whether he had thought of taking the ideas in the film and translating them into a quantitative model. Anfinsen said that he did not really know how one would do this, but to me it suggested an approach to the mechanism of protein folding, at least for helical proteins such as myoglobin.

When David Weaver joined my group at Harvard, while on a sabbatical leave from Tufts, we developed what is now known as the diffusion–collision model for protein folding [**Karplus and Weaver, 1976; Karplus and Weaver, 1994**]. Although it is a simplified coarse-grained description of the folding process, it showed how the search problem for the native state could be solved by a divide-and-conquer strategy. In addition to providing a conceptional answer to the Levinthal paradox, the diffusion–collision model made possible the estimation of folding times, something not available from schematic descriptions of protein-folding mechanisms. The model was ahead of its time because data to test it were not available. It was only a decade later that experimental studies demonstrated

———— ❦ ————

[1] Of course, it is now known that some proteins, such as the supramolecular complex GroEL, have more complex folding mechanisms and require chaperones to fold. GroEL is a "molecular machine" for which we helped to elucidate the mechanism by use of molecular dynamics simulations [**Ma et al., 2000; van der Vaart, Ma, and Karplus, 2004**].

that the diffusion–collision model describes the folding mechanism of many helical proteins [**Islam, Karplus, and Weaver, 2002**] as well as some others [**Islam, Karplus, and Weaver, 2004**].

During my visit to Lifson's group, I learned about their work on developing empirical potential energy functions. The novel idea was to use a functional form that could serve not only for calculating vibrational frequencies, as did the expansions of the potential about a known or assumed minimum-energy structure, but also for determining that structure. The "consistent force field" (CCF), introduced by Lifson and his coworkers, included nonbonded interaction terms, so that the minimum-energy structure could be found after the energy terms had been appropriately calibrated [**Lifson and Warshel, 1969**]. The possibility of using such energy functions for larger systems struck me as potentially very important for understanding biological macromolecules such as proteins, though I did not begin working on this immediately.

Once Attila Szabo had finished the statistical mechanical model of hemoglobin cooperativity, I realized that his work raised a number of questions that could be explored only with a method for calculating the energy of hemoglobin as a function of the positions of its constituent atoms. However, at this time, I was not aware of any method for doing such a calculation. Bruce Gelin had begun theoretical research in my group as a graduate student in 1967. He started out by studying the application of the random-phase approximation to two-electron systems, such as the helium atom. This was the Vietnam War era and after two years at Harvard, Gelin was drafted. He was assigned to the military police in a laboratory concerned with drug usage (including lysergic acid diethylamide [LSD]). Paradoxically, this work aroused his interest in biology, and when he returned to finish his degree Gelin wanted to change his area of research to a biologically related problem.

The time was ripe to develop a program that would make it possible to take a given amino acid sequence (e.g., that of the hemoglobin α chain) and a set of coordinates (e.g., those obtained from the X-ray structure of deoxy hemoglobin) and to use this information to calculate the energy of the system and its derivatives as a function of the atomic positions. This could be used for perturbing the structure (e.g., by binding oxygen to the heme group) and finding a new structure by minimizing the energy. Developing the program would be a major task, but Gelin had the right combination of abilities to carry it out [**Gelin, 1976**].

The program Gelin developed, while not trivial to use, was applied to a variety of problems, including Gelin's pioneering study of aromatic ring flips in the bovine pancreatic trypsin inhibitor (BPTI) [**Gelin and Karplus, 1975**] as well as his primary project on hemoglobin. The idea was to introduce the effect of ligand binding on the heme group as a perturbation (undoming of the heme) and to use energy minimization to determine the response of the protein to the perturbation. Because computing at the Harvard Computer Center was too expensive, an IBM 7090 at Columbia University was our workhorse at the time; the Columbia computing staff continued to give me access even though I had moved to Harvard. Since nothing like this calculation had been done before, it required considerable courage to attempt it with the available computers, not the least element being that Gelin's PhD depended on it.

Gelin's efforts were successful! His work introduced a new dimension in theoretical approaches to understanding protein structure and function. Gelin showed how the effect of undoming of the heme induced by the binding of oxygen was transmitted to the interface between the hemoglobin subunits. The analysis provided an essential element in the cooperative mechanism in its demonstration at an atomic level of detail how communication between the subunits occurred [**Gelin and Karplus, 1977**]. Another early application of the program was Dave Case's simulation of ligand escape after photodissociation from myoglobin [**Case and Karplus, 1979**], a study that was followed by the work of Ron Elber [**Elber and Karplus, 1990**], which gave rise to the locally enhanced sampling (LES) and multiple copy simultaneous search (MCSS) methods. The latter was developed a few years later by Andrew Miranker as a fragment-based approach to drug design [**Miranker and Karplus, 1991**].

Gelin would have faced an almost insurmountable task in developing the program, which we now refer to as pre-CHARMM, if there had not been prior work by others on protein energy calculations. Although many persons contributed to the development of empirical potentials, the two major inputs to our work came from Lifson's group at the Weizmann Institute and Scheraga's group at Cornell University [**Scheraga, 1968**]. When Warshel had come to Harvard, he had brought the Consistent Force Field (CFF) program with him. His presence and the availability of the CFF program were an important resource for Gelin, who was also aware of Michael Levitt's pioneering calculations of proteins [**Levitt and Lifson, 1969**].

The program has been considerably restructured and has continued to evolve over the intervening years. In preparing to publish a paper on the program in the early 1980s [**Brooks *et al.*, 1983**], mainly to give credit to the important contributors, we felt we needed a name. Bob Bruccoleri came up with HARMM (HARvard Macromolecular Mechanics), which seemed not the ideal choice. However, Bob's suggestion inspired the addition of a "C" for chemistry, resulting in the name CHARMM. I sometimes wonder if Bruccoleri's original suggestion would have served as a useful warning to inexperienced scientists working with the program. If one did not understand the limitations of the program, particularly the approximate energy function employed at the time, it was unlikely that attempts to use the program would lead to meaningful results.

The present-day CHARMM program is a great advance from the original version. It is described in a 2009 paper [**Brooks *et al.*, 2009**], which lists many of the contributors, who are part of the group of CHARMM developers, as they are now called. Most of them had been students or postdoctoral fellows in my group.[2] The CHARMM program is distributed worldwide in both academic and commercial settings.

Figure 14.1. CHARMM developer meeting at Harvard, 2017 (photo by Wonmuk Hwang)

[2]There is an annual three-day meeting where the developers report on what they have accomplished during the year (Figure 14.1). The meeting, which was originally held only at Harvard, now takes place in a variety of places in the United States and Europe, usually hosted by one of the "core" developers. There are now almost forty developers who participate. An essential part of the meeting is a banquet in a carefully selected restaurant.

Chapter 15

The First Molecular Dynamics Simulation of a Biomolecule

———————— ⊙⟊⊙ ————————

G iven that pre-CHARMM could calculate the forces on the atoms of a protein, the next step was to use these forces in Newton's equation to calculate the dynamics. This fundamental development was introduced in the mid-1970s when Andy McCammon joined my group. An essential element that encouraged us in this attempt was the existence of molecular dynamics simulation methods for simpler systems. Molecular dynamics had followed two pathways, which come together in the study of biomolecule dynamics. One pathway concerns trajectory calculations for simple chemical reactions. My own research on the H + H_2 reaction [**Karplus, Porter, and Sharma, 1965**] had served as preparation for the many-article problem posed by biomolecules. The other pathway in molecular dynamics concerns physical rather than chemical interactions and the thermodynamic and dynamic properties of large numbers of particles rather than detailed trajectories of a few particles. Although the basic ideas go back to van der Waals and Boltzmann, the modern era began with the work of Alder and Wainwright [**Alder and Wainwright, 1957**] on hard-sphere liquids in the late 1950s. The paper by Rahman [**Rahman, 1964**] describing a molecular dynamics simulation of liquid argon with a soft-sphere (Lennard-Jones) potential represented an essential next step. Simulations of more complex fluids followed; the now classic study of liquid water by Stillinger and Rahman was published in 1974 [**Stillinger and Rahman, 1974**], shortly before our protein dynamics simulations.

The background I have outlined set the stage for the molecular dynamics simulations of biomolecules. The size of an individual protein molecule, composed of 500 or more atoms for even a small protein, is such that its simulation in isolation can serve to obtain approximate equilibrium properties, as in the molecular dynamics of fluids. Concomitantly, detailed aspects of the atomic motions are of considerable interest, as in trajectory calculations. A basic assumption in initiating such studies was that potential functions could be constructed, which were sufficiently accurate to give meaningful results for systems as complex as proteins or nucleic acids. In addition, it was necessary to assume that for these inhomogeneous systems, in contrast to the homogeneous character of even complex liquids such as water, simulations of an attainable timescale (10–100 ps) could provide a useful sample of the phase space in the neighborhood of the native structure. There was no compelling evidence for either assumption in the early 1970s. When I discussed my plans with chemistry colleagues, they thought such calculations would be meaningless, given the difficulty of treating few atom systems accurately; biology colleagues felt that even if we could do such calculations, they would be a waste of time. By contrast, the importance of molecular dynamics simulations in biology was supported by Richard Feynman's prescient statement in the well-known textbooks based on his physics lectures at Caltech:

> Certainly no subject or field is making more progress on so many fronts at the present moment, than biology, and if we were to name the most powerful assumption of all, which leads one on and on in an attempt to understand life, *it is that all things are made of atoms* [italics in the original], and that *everything that living things do can be understood in terms of the jigglings and wigglings of atoms.* [italics added] [**Feynman, Leighton, and Sands, 1963**].

The original simulation, which was published in 1977 [**McCammon, Gelin, and Karplus, 1977**], concerned the bovine pancreatic trypsin inhibitor (BPTI), which has served as the "hydrogen molecule" of protein dynamics because of its small size, high stability, and an accurate X-ray structure that was available in 1975 [**Deisenhofer and Steigemann, 1975**]. BPTI could be obtained in large quantities because it was used as a drug, Aprotinin, to slow down fibrinolysis that leads to the break up of blood clots in operations; its inhibition of trypsin may not have a biological function. This is an illustration of the fact that when a protein is observed to act in a given way (e.g., BPTI was found to bind to trypsin), this may have nothing to do with its biological role.

In the mid-1970s, it was difficult to obtain the computer time required to do such simulations in the United States. However, the Centre Européen Calcul Atomique et Moléculaire (CECAM), located outside of Paris, had access to a large computer that was available for scientific research. (Equivalent computers in the United States were found only in the defense agencies and were not generally available to university researchers.) CECAM's founding director and guiding spirit was Carl Moser. A CECAM workshop (a two-month *workshop* worthy of its name) was organized there by Herman Berendsen in 1976 with the title "Models for Protein

Dynamics." His proposal for the workshop states: "The simulation of water was a first topic to be studied. The application to proteins was then not foreseen in five or ten years to come." By this, he implied that he thought protein simulations were not actually going to be a subject of this workshop. Thus, I am sure he was surprised that Andy McCammon did the first protein simulation *during* the workshop.

Realizing that the workshop was a great opportunity to do the required calculations, Andy McCammon and Bruce Gelin worked very hard at Harvard before the workshop to prepare and test a program for doing the molecular dynamics simulation of BPTI. Once we arrived at CECAM, Andy was able to do the simulation on its IBM 137/168, a "supercomputer" of that time. It had a top speed of a million floating point operations (flops) per seconds.

The discussions during the workshop were important in introducing many others who became active in the field (including Herman Berendsen, Wilfred van Gunsteren, Michael Levitt, and Jan Hermans) to the possibility of doing such calculations [**Berendsen, 1976**].[1]

Although the original simulation was done in vacuum with a crude molecular mechanics potential and lasted for only 9.2 ps (Figure 15.1), the results were instrumental in replacing the view of proteins as relatively rigid structures (in 1981, Sir D.L. Phillips commented, "Brass models of DNA and a variety of proteins dominated the scene and much of the thinking" [**Phillips, 1981**]) with the realization that they were dynamic systems whose internal motions could play a functional role. Of course, there already existed some experimental data, such as the hydrogen exchange experiments of Linderstrom-Lang and his coworkers [**Hvidt and Nielsen, 1966; Linderstrom-Lang, 1955**], pointing in this direction. It is now recognized that the X-ray structure of a protein provides

---

[1]The meeting organized in 2016 by Emanuele Paci with Dominic Tildeley and Benoit Roux, to commemorate the 40th anniversary of the 1976 CECAM workshop on "Models of Protein Dynamics" prompts me to include a footnote here, based on my unpublished biography of Carl Moser, which was presented as part of the workshop program. In the biography, I describe how Carl became dissatisfied with his research in quantum chemistry and decided to focus his talents to create CECAM and its workshops. He did this, as he told me, because he had concluded that his research would "never lead to a Nobel Prize." However, the simulation was performed on the IBM 137/168 with a top speed of million floating operations (flops) per second. Carl Moser, CECAM, and the Orsay Computer Center played a significant role in making possible the first molecular dynamics simulation of a protein, an essential element in the 2013 Nobel Prize in Chemistry, though not in the Nobel citation. Over the years, ever faster computers have been built. Making use of multiprocessor GPU/CPU systems, petaflop and even exaflops computers are becoming available.

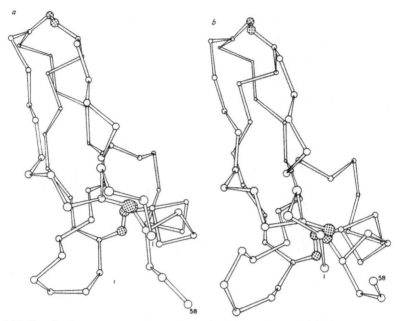

Figure 15.1. Two bovine pancreatic trypsin inhibitor (BPTI) structures from the simulations. The peptide backbone carbons are shown as open circles and the disulfide bonds are shown between stippled atoms. (a) X-ray structure. (b) Time evolved structure after 3.2 ps of molecular dynamics simulation [Drawn by Bruce Gelin and is reproduced with permission from McCammon, Gelin, and Karplus, 1977]

the average atomic positions but that the atoms exhibit fluid-like motions of sizable amplitudes with respect to these averages. Protein dynamics subsumes the static picture. The average positions are essential for the discussion of many aspects of biomolecular function in the language of structural chemistry, but the recognition of the importance of fluctuations opened the way for more sophisticated and accurate interpretations of functional properties.

The conceptual changes resulting from the early studies make one marvel at how much of great interest could be learned with so little—such poor potentials, such small systems, so little computer time. This is, of course, one of the great benefits of taking the initial, somewhat faltering steps in a new field in which the questions are qualitative rather than quantitative and any insights, even if crude, are better than none at all.

Chapter 16

Early Applications of Molecular Dynamics

―――――⊙⊙――――――

Molecular dynamics simulations of proteins and nucleic acids, as of many other systems composed of particles (e.g., liquids, galaxies), can in principle provide the ultimate details of motional phenomena. The primary limitation of simulation methods is that they are approximate. Here, experiment plays an essential role in validating the simulation methods; that is, comparisons with experimental data serve to test the accuracy of the calculated results and provide criteria for improving the methodology. Although the statistical errors can be calculated [**Yang, Bitetti-Putzer, and Karplus, 2004**], estimates of the systematic errors inherent in the simulations have not been possible; for example, the errors introduced by the use of empirical potentials are difficult to quantify. When experimental comparisons indicate that the simulations are meaningful, their capacity for providing detailed results often makes it possible to examine specific aspects of the atomic motions far more easily than by using laboratory measurements.

Two years after the bovine pancreatic trypsin inhibitor (BPTI) simulation, it was recognized [**Artymiuk** *et al.*, **1979; Frauenfelder, Petsko, and Tsernoglou, 1979**] that thermal (B) factors determined in X-ray crystallographic refinement could provide information about the internal motions of proteins. Plots of estimated mean-square fluctuations versus residue number (introduced in the original BPTI paper [**McCammon, Gelin, and Karplus, 1977**]) have become a standard part of papers on high-resolution structures, even though the contribution to the B factors of overall translation and rotation and crystal disorder persist as a concern in their interpretation [**Kuriyan and Weis, 1991**]. During the decade following the first simulation, a range of phenomena were investigated by molecular dynamics simulations of proteins and nucleic acids. It is fair to say that a plethora of experimental data were just waiting for molecular dynamics simulations to elucidate them. Many of these early studies were made by my students at Harvard and focused on the physical aspects of the internal motions and the interpretation of experiments. They include the analysis of fluorescence depolarization of tryptophan residues [**Ichiye and Karplus, 1983**], the role of dynamics in measured nuclear magnetic resonance (NMR) parameters [**Dobson and Karplus, 1986; Levy, Karplus, and Wolynes, 1981; Olejniczak** *et al.*, **1984**] and inelastic neutron scattering [**Cusack** *et al.*, **1986; Smith** *et al.*, **1986**], and the effect of solvent and temperature on protein structure and dynamics [**Brünger, Brooks, and Karplus, 1985; Frauenfelder** *et al.*, **1987; Nadler** *et al.*, **1987**]. The now widely used simulated annealing methods for X-ray structure refinement [**Brünger and Karplus, 1991; Brünger, Kuriyan, and Karplus, 1987**] and NMR structure determination [**Brünger** *et al.*, **1986; Nilsson** *et al.*, **1986**] also originated in this period. Simultaneously, a number of applications demonstrated the importance of internal motions in biomolecular function, including the hinge-bending modes for opening and closing active sites [**Brooks and Karplus, 1985; Colonna-Cesari** *et al.*, **1986**], the flexibility of tRNA [**Harvey** *et al.*, **1984**], the induced conformation change in the activation of trypsin [**Brünger, Huber, and Karplus, 1987**], the fluctuations required for ligand entrance and exit in heme proteins [**Case and Karplus, 1979; Elber and Karplus, 1990**], and the role of configurational entropy in the stability of proteins and nucleic acids [**Brooks and Karplus, 1983; Irikura** *et al.*, **1985**]. Many of these studies, which were done in the early 1980s, seem to have been forgotten. In any case, they are only rarely cited in the current literature. Of course, when the studies are redone with the more accurate potential functions now available and much longer simulation times now possible (nanoseconds instead of picoseconds), they yield improved results, though most often they confirm the earlier work.

Given the recognition of the importance of the field of molecular dynamics simulations, many departments were interested in having a practitioner on their faculty and it was relatively easy for my graduate students and postdoctoral fellows to obtain positions in good universities. However, as time went on, a saturation effect took place and obtaining an academic job became more difficult. Good publications, *per se*, were not sufficient and I realized that a paper in a journal such as *Science* or *Nature* was important. One of my outstanding postdocs, Yaoqi Zhou, was having a difficult time in finding a position, in spite of several very good publications. One of these concerned the dimeric hemoglobin of *Scapharca,* a marine mollusc [**Zhou and Karplus, 2003**], in which each of the subunits is very similar in structure to that of human hemoglobin, but they are arranged very differently. It is an interesting question, how the human hemoglobin tetramer with its higher cooperativity evolved from a dimeric precursor.

Yaoqi had also done a pioneering study of a three-helix bundle protein with a C_α model for the protein chain and a square-well interaction potential for pairs of nonbonded residues. These simplifications made possible the use of a discrete molecular dynamics algorithm for studying the folding process [**Zhou and Karplus, 1999**]. The speed of the latter is such that several hundred folding trajectories could be calculated for different relative weights of native and non-native interactions in the model potential function. Figure 16.1 shows a typical trajectory for a model (a) with the native state strongly favored (Go-like potential) and (b) in which nonnative interactions make a significant contribution along the folding pathway. In the former model, the helices form first and then diffuse

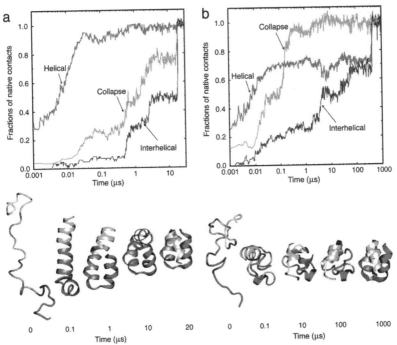

Figure 16.1. Simulations of the folding of a three-helix bundle protein. (Upper) (a and b) Semilog plots of the time dependence of the fractions of native helical and interhelical contacts and the inverse fractions of native volume (calculated from the inverse cube of the radius of gyration) for two different trajectories. (Lower) Structures of the protein molecule at selected times [Reproduced with permission from Zhou and Karplus, 1999 (Copyright 1999, Nature Publishing Group)]

to find the native fold (a limiting case of the diffusion–collision model [**Islam, Karplus, and Weaver, 2002**]), whereas in the latter model, there first is a collapse to a relatively disordered globule and the helices form simultaneously with the native tertiary structure. I thought this paper was of sufficient interest to be published in a "high-impact" journal and submitted it to *Nature*, which accepted it [**Zhou and Karplus, 1999**] without any of the back-and-forth discussion required for publication of the 11-*cis* retinal paper (see Chapter 12). Once the paper was published, Yaoqi received several job interviews, and he accepted an offer from the State University of New York in Buffalo.

Chapter 17

My Career as a Photographer

------- ❦ -------

When I completed my PhD, my family gave me a Leica IIIC, a superb camera, which my Uncle Alex had brought to the United States from Vienna. Throughout my travels as a postdoctoral fellow in Europe and since then, I took photographs, particularly of people, using a trick I had developed. The Leica had a long focus (Hector) lens with a reflex viewer, which enabled me to face away from my subject when I took their photograph. This made it possible for me to take photographs of crowds and individuals without their being aware of it. Often it is clear from looking at the photograph that the subjects do not know that I am taking a photograph, but that they are trying to see what I am photographing (Figure 17.1).[1]

Figure 17.1. Navaho family near Window Rock, Arizona, 1956

------- ❦ -------

[1] I later learned that a number of other photographers had used a similar technique for their street photography. One of them is Paul Strand (1890–1976), who is famous for his black-and-white photographs. As a street photographer he wore a heavy overcoat in which he could hide his camera. Ironically, there is his iconic 1916 picture of a beggar woman taken while she wears a sign saying "I AM BLIND."

Figure 17.2. Looking toward Castle Peak from Kowloon[2]

There were few scientific interactions during my extended trips in Europe and North America, but I learned much about peoples and their cultures, art, architecture, and cuisine, all of which continue to play a role in my life. Some of the later extended trips were made possible by scientific meetings in South America and Asia to where I was invited and my travel expenses were paid by the organizers. One of these was to Japan from where I took a side trip to Hong Kong and Kowloon (Figure 17.2).

Another trip, which enabled me to visit many part of Brazil (Figures 17.3 and 17.4), was organized by my friend Aron Kupperman.[3]

Most of the European trips were made in a Volkswagen Beetle (Figure 17.5). I had purchased the Beetle jointly with Gary Felsenfeld. We had met when we both were undergraduates at Harvard, then we had both gone to Caltech for graduate study. Gary, who had been working primarily with Verner Schomaker,

---

[2] This photograph, which was taken in 1962, shows a view of Hong Kong, partly hidden behind the Chinese "dragon boats." One can see that Hong Kong was a village at the time. The same view today would be dominated by skyscrapers.

[3] Aron arranged for me to be invited to make a survey of chemistry research being done in Brazil. The purpose from our point of view was to make it possible for me to travel throughout Brazil, adding photographs to my collection. However, I did make the survey and submitted a report to the office of President Kubitschek. By 1962 when I submitted the report, he was already no longer president, his term having ended in 1961. Selma Jeronimo, a Brazilian friend of my daughter Tammy, recently found a record of my report in the presidential archives, but not the actual report.

Figure 17.3. Overlooking Rio de Janeiro, 1960

Figure 17.4. Brazilia, 1960

Figure 17.5. Reflection in a MG hub cap, while waiting for the ferry to Denmark, 1954

received his PhD a year after I did and then also came to Oxford to work with Coulson. Some of the trips were taken jointly with Gary and sometimes I was by myself. Being by myself was actually an advantage because it made it possible for me to wait, sometimes for many minutes, for the "decisive moment" in the words of Henri-Cartier Bresson.

I took several thousand Kodachrome slides with my Leica IIIC primarily during my travels in the 1950s and 1960s. Once Reba and Tammy were born, I mainly photographed my children as they grew up, sometimes in interesting places as we did not stop traveling. During the years following the 1950s, most of the Kodachromes were kept in metal boxes, though some of the ones from the 1970s were still in the original yellow boxes in which they were returned after being developed. We rarely projected them. This turned out to be important for their preservation because although Kodachrome images are stable in the dark, they are sensitive to light, particularly to the intense light used in projectors.

When I had been unable to obtain Kodachrome film while in England, I had taken pictures with Ektachrome film from Kodak. I looked at them some years ago and found that the colors had faded to a pink hue, although the image was still there.

Once we were settled in Strasbourg (see Chapter 18), Marci and I spent many evenings looking through the slides from the 1950s and 1960s and choosing the better slides. In most cases the Kodachromes had been perfectly preserved, so that the colors of the slides were unchanged. In a few cases, bacteria had attacked the protective gelatin, so that large uniform areas (e.g., blue sky) had spots that had to be repaired.

Marci felt that the slides should be printed and had investigated in both Boston and Strasbourg to find someone or a workshop that would reliably do this. She had not come up with anyone. Then, while we were in Oxford during the academic year 1999–2000 when I was Eastman Professor, Marci had the luck to be introduced to a marvelous photographic craftsman, Paul Sims (Colourbox Technique). Initially, Paul produced standard size prints using an old-fashioned printer, which he kept functioning by ordering replacement parts from other printers that were no longer in use. He made beautiful four-by-six inch photographic prints, which Marci carefully collected in several photo albums, which are now with us in Cambridge.

I had no idea that my pictures were anything special. However, as Paul became familiar with the collection of photographs, he told Marci that they were outstanding and should be exhibited. At this point I started to work with Paul, who had begun to shift his efforts to digitizing the Kodachrome transparencies. He made a set of 67 beautiful exhibition prints on archival paper (printed to last about 65 years) from some of the slides, which formed my first public exhibition. It took place in Wolfson College, Oxford, in November 2006. A selection was also exhibited at my 75th birthday celebration at NIH in Bethesda, Maryland [**Post and Dobson, 2005**].

Soon afterwards exhibitions were held in a number of venues.[4] However, it was the one at the Bibliothèque Nationale de France (BnF) in Paris that paved the way for many others. The exhibition took place in the spring and summer of 2013. I am particularly proud that this exhibition was organized before I received my Nobel Prize that fall. That it took place at all is one of the many serendipitous events in my life.

---

[4]See Appendix 2.

I was collaborating with Jean-Pierre Changeux at the Institute Pasteur in Paris on applying simulation methods to interpret the function of the nicotinic receptor [**Taly et al., 2006**]. I knew that Changeux was interested in art and, in fact, had a large personal collection of 16th- and early 17th-century French paintings, which he had donated to the Musée Bossuet. After we had finished discussing science, I "shyly" (as he describes it) took out a book that Paul Sims and I had prepared showing a set of the Kodachrome photographs [**Martin Karplus, *Images from the 50's* (Blurb.com, 2011)**].[5] To my pleasant surprise, Changeux was very impressed by them and said he would contact Sylvie Aubenas, Directeur du département des Estampes et de la Photographie, at the BnF. He knew her well because he was a member of the BnF committee that decided which proposed donations of photographs the BnF should accept; one example is the collection of Brassaï.

When Sylvie Aubenas saw the photographs on my computer, she was very enthusiastic about them, and I arranged to come from Strasbourg to show her some of the actual exhibition prints. This was in 2011. Once she had seen them, she decided to propose an exhibition at the BnF. It required considerable determination on her part, as well as support from Changeux, to make it happen. In particular, she had to convince Bruno Racine, the president of the BnF. The BnF is perhaps the most prestigious photographic gallery in France and it had never had an exhibition of an "unknown" photographer, like myself. Moreover, the BnF in Rue Richelieux, where photographic exhibitions had been held, was closed for renovation. So it was decided to hold it at the new library, the Bibliothèque Mitterand, the building of which was started during his presidency but only finished in 1996.

The exhibition was finally entered in the BnF calendar for May 2 through August 15, 2013 in the Allée Julian Cain (Figure 17.6).[6]

The Allée is a wonderful space with excellent natural lighting. Moreover, it connects two parts of the library so that many people visiting or working at the library walked by, whether they were interested in seeing the exhibition or not.

---

[5]Blurb is a website that played an important role in making it possible for a photographer to publish a photobook without having to advance a large amount of money. Basically, at that time, you had to pay about $100 per book and order a minimum of two copies. Then Blurb would print additional "copies on demand" with each copy also costing about $100. This is significantly more per copy than if the book were published with a print run of 1,000.

[6]Julian Cain had been the administrator of the BnF before the war, had survived Buchenwald, and resumed his position until 1964. He had managed to hide many of the BnF's most valuable items from the Nazis.

Figure 17.6. Entrance to the Exhibition at the Bibliothèque Nationale de France

In addition, the exhibition had been well advertised, so many people, including tourists, came to see it.[7]

There was space for about 100 photographs. Of the approximately 450 that I had selected from the 4,000 or so Kodachromes I had taken in my travels between 1953 and 1965, I finally chose about 200 to show to Sylvie on my computer at our initial meeting to select the photographs for the exhibition. After she had had a chance to think about the images, she and I met again to choose the 100 for the exhibition. We agreed on most of them, but there was "horse trading" on some. The picture of the Iguazu Falls between Brazil and Argentina (Figure 17.7) was one that Sylvie did not include in her list to show, but with the help of Valerie Prevot, who was in charge of the actual installation of the exhibition, it was finally included. Sylvie felt that the exhibition should be focused on my pictures of people, which certainly were my primary interest. Still, the Iguazu Falls "landscape" image was so special that I thought it should be in the exhibition.

---

[7]Philippe Meyer, a reporter for *France Culture*, devoted one of his "Chronique" to the exhibition. Each morning at about 7:50 am, he described a cultural event in four minutes. Many people listened to him while having their morning coffee to find out what was interesting to do that day. He described my exhibition, in part saying, "There can be found one hundred or so photos which show, with variant diversity, the palette of the colors of the 1950s. Their charm is immediate, their eloquence, if one can speak of images being eloquent, is overwhelming." (Translation from the French by Mary Podevin and Mischa Karplus).

Figure 17.7. Iquazu Falls, Brazil/Argentina, 1960

Rather than using the prints prepared by Paul Sims, the BnF had new prints made from the digitized images by the Laboratoire Picto in Paris under Sylvie's supervision.

One important part of the exhibition involved photographs from Yugoslavia. When I took the photographs in 1954, Yugoslavia was one country and was actually very peaceful with Serbs, Albanians, Croatians, and Macedonians living together as did people with different religions, Muslims, Christians, and Jews. Tito's death in 1980 eventually led to the disintegration of Yugoslavia. The relationships among these groups rapidly broke down and much fighting and destruction occurred. One thing I had to do for the exhibition at the BnF (and the subsequent exhibitions) was to figure out where I actually had taken the photographs, whether it was North Macedonia (the official name), Croatia Herzegovina, etc. By making a list of the towns where the pictures had been taken, I was able to figure out in which country they now were in. Very helpful also were the detailed letters I had written to my parents about my trips (Figure 17.8).

On May 14, 2013, Marci and I were at the BnF for the opening. It was a grand affair with many friends from Paris and elsewhere (Figures 17.9 a–c).

Figure 17.8. Fisherman on Lake Vransko, Croatia, 1954

Figure 17.9a. Bruno Racine and Bernard Bigot with me

Figure 17.9b. Jean-Pierre Changeux, unknown individual, Bruno Racine, and Sylvie Aubenas with me

Sadly Paul Sims had passed away earlier in the year, but his widow Joey and their daughter were there. My children were not able to participate at that time of year, so it was arranged that we would meet in Paris during the summer vacation. Mischa and I flew in from Boston, while Reba, Tammy, and her daughter Rachel arrived in Paris on the way to our annual summer vacation at the chalet in the Haute Savoie (see Chapter 13).

As it happened, the Hotel des Deux Continents had an anniversary celebration and reduced prices for old-time clients. Certainly I qualified as an old client from my first trip to Paris in 1953 when Sidney Bernhard had introduced me to the hotel. So we decided to stay there while in Paris although Marci and I now usually stayed in different hotels. I was sitting in the lobby waiting for the children to come down when who should I see but Gary Felsenfeld and his wife Naomi. On parts of the trip during which I had taken the photographs in what was then Yugoslavia, Gary and I had driven together in the Volkswagen Beetle, which we had purchased jointly. They went to the exhibition at the BnF and realized that Gary had been with me when some of the pictures were taken, so it was a poignant get together.

One of the biggest changes in my life resulting from the Nobel Prize was that I received more invitations to exhibit my photographs. The first of these took

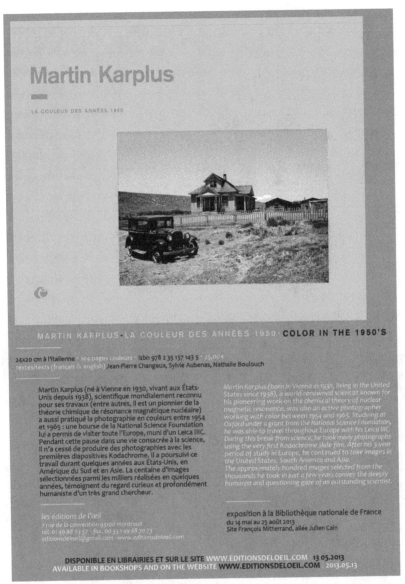

Figure 17.9c. Exhibition catalog

place in New York at the Austrian Cultural Forum. Before deciding whether to accept, Marci and I visited the gallery in its building, designed by the Austrian architect Raimund Abraham. It turned out to be very narrow with essentially one exhibition room per floor. I was rather skeptical about showing the photographs there. However, in collaboration with the Director Christine Moser and Natascha

Boojar, who was in charge of exhibitions, we were able to use the floors in a creative way. It turned out to be what I think was (and remains) one of the best exhibition of my photographs. The exhibition was also the first in which I showed some more recent photographs from India and China (Figures 17.10 and 17.11) taken with my first digital camera, the Canon E0S 30D, which had replaced my Leica.[8] The camera was a gift from my students at the 75th birthday celebration held at NIH in 2005.

The exhibition was opened on September 24, 2014 (Figures 17.12 and 17.13), by Heinz Fischer, President of Austria at the time; he was in New York for the General Assembly of the United Nations. He was aware of the remarks I made shortly after the announcement of the Nobel Prize in an interview with ORF, the Austrian public radio station, concerning the anti-Semitism that I had observed in Vienna during a visit in 1998.

Figure 17.10. Chandigarh Rock Garden, India, 2009

[8]Some of the photographs shown there were shipped from Strasbourg, where there is a permanent exhibition of a selection of my photographs in the atrium of ISIS, the Institute at the Université de Strasbourg, where I am Professeur Conventionée.

Figure 17.11. In a Naxi market, Lijiang, China, 2008

Marci and I had been in Vienna at a scientific meeting organized by Peter Schuster, a professor at the University of Vienna, and we stayed at a small pension at the edge of the central part of the city. It was November 1998, the 60th anniversary of Kristallnacht (Night of Broken Glass), when Nazi SS and storm troopers, throughout major cities in Germany and Austria, smashed the store fronts of Jewish owned shops, set fire to synagogues, and accosted Jews. The city of Vienna commemorated the anniversary by having photographs of what had happened at each location on kiosks throughout the city. I noticed that evening that one

Figure 17.12. Poster at Austrian Cultural Forum exhibition at entrance

such kiosk, which was visible from our pension, had been covered with swastikas and anti-Semitic slogans. I phoned the city office responsible for the commemoration. The woman I spoke with said yes, she was aware of this and that every morning, clean photographs were put up to replace the defaced photographs. When I inquired why they did not try to arrest the hooligans involved, she hung up.

Figure 17.13. Marci, Mischa, and I at the Austrian Cultural Forum exhibition opening

At one point during our stay, Marci asked the receptionist at the pension for information about how to get to the Karplusgasse, which we knew was a street on the outskirts of Vienna that had been named in honor of my grandfather John Paul Karplus (see Chapter 1). The pension clerk looked at her and said something like, "I cannot understand why one would name a street after a ... ur, Jewish doctor," somewhat stumbling over her words to avoid being blatant.

In opening the exhibition, President Fischer remarked that Austria was very different now as I would discover when I came to Vienna to receive an honorary doctorate from the university. My response, thanking him for his kind remarks, was that I was hopeful that he was correct but, as a scientist, I would make my own observation when I was there.

The first of several exhibitions of the BnF collection after the original one in Paris took place in Berlin in November 2014. This was thanks to Peter Badge, who was taking photographs of all the Nobel Prize winners as one of his major projects. Each time the Nobel Prize winners were announced, he traveled to visit them to take a portrait. In my case, Badge had come to our house in Cambridge. As we were talking, I discovered that he was a good friend of the owner of the Einstein Café and Gallery in Berlin. He suggested arranging for the BnF exhibition to be shown there. Since there was room for only about half of the total BnF set, the number of photographs had to be pared down and then rearranged in a creative way; the resulting exhibition turned out very well.

Since the exhibition was in Berlin, the Minister of Culture, Professor Monika Grütters, came to the opening and presented introductory remarks that presaged those made the following year in the ceremonies in Vienna. She remarked that it was far from certain, given what had happened to my family and my experience as a child, that I would be willing to have an exhibition in Berlin. Further, she said that she regarded this exhibition as both an admonition to prevent anti-Semitism and a gift because the exhibition made it possible for Germany to demonstrate that it has learned from its past.

I have continued to take photographs during my travels (see Appendix 2), the most recent being in Israel, Cuba, Morocco, and Tibet (Figures 17.14a, 17.14b through 17.17). A picture of me with my Canon camera is shown in Figure 17.18.

Figure 17.14a. Israel/Palestine Wall, Palestinian side, 2014

Figure 17.14b. Israel/Palestine Wall, Palestinian side, 2014

Figure 17.15. Havana, Cuba, 2015

Figure 17.16. Marakesh, Morocco, 2015

Figure 17.17. Near Lhasa, Tibet, 2015

Figure 17.18. Photograph of me with my digital Canon camera

Chapter 18

How We Came to Move to Strasbourg

————— ⟡ —————

Since the 1970s, my frequent sojourns in France were spent primarily in Paris, on the one hand, and at our chalet in Manigod in the Haute Savoie, on the other. However, this was changed by a serendipitous event.

In the summer of 1991, Jean-François Levefre and Oleg Jardetsky organized the first International School of Biological Magnetic Resonance in Erice, Italy, at the Ettore Majorana Center. Given my work on the Karplus equation, Jean-François, whom I had met previously at a conference, invited me to be one of the keynote lecturers. The center had been set up in honor of Ettore Majorana, a brilliant young physicist who was one of Enrico Fermi's best students. He was born in 1906 in Catania, Sicily, and disappeared under mysterious circumstances in 1938.

The international schools at Erice are very unusual in that the whole village participates in them. Erice is on a plateau in Sicily near Catania and sufficiently high to be pleasant, even in the summer when much of Sicily is very hot. The main meetings take place in a disaffected church with excellent acoustics. All the restaurants in Erice are open to the conferees, who receive tickets for their dinners and can choose in which restaurant to dine. This resulted in small groups eating together to discuss the topics of the day, or anything else for that matter. The food might not be three-star, but the warm atmosphere is something that I have never seen at any other conference. Moreover, an important aspect of the school is that about thirty or so postdoctoral students are invited and these informal arrangements lead to relaxed interactions between the lecturers and students.[1]

There was a small mansion in Erice where the mayor had once lived. He had moved to a larger house and the mansion was now reserved for use by the lecturers. It is situated on a hill above the plateau, with a beautiful view. In the relaxed atmosphere of the meeting, while working in the mansion I finally solved a problem that I had been working on unsuccessfully for sometime. It involved understanding the significance of the results of free energy simulations by their

————— ⟡ —————

[1] One day Enrico Clementi, the quantum chemist, who was also at the conference, reported that his wallet had been stolen. That evening the wallet was returned. We were told that the town was under the control of the Mafia, which was very proud of it, and so anything like this would immediately be taken care of. Exactly what "taken care of" meant, other than returning the wallet, we never found out.

decomposition into component contributions. My interest in this decomposition, as I already mentioned in relation to other studies, was to be able to use the calculations to deepen our understanding. Such an approach, even if approximate, is of interest because it provides information about the origin of the difference in free energy behind two states, rather than just a single number.

I had proposed this approach sometime before, fully realizing that in such a decomposition one was not dealing with exact thermodynamic quantities. This problem with the method had also been pointed out by van Gunsteren [**Mark and van Gunsteren, 1994**]. I had been trying to develop a theoretical justification for my belief that the decomposition was meaningful but had had trouble in getting to the final stage of the argument. In Erice, I was able to show that the decomposition is exact in lowest order and that there are higher order corrections. The analysis and results are given in a paper I published with Stefan Boresch [**Boresch and Karplus, 1995**].

During the Erice meeting, Marci and I often dined with Jean-François and his wife Christine. There was what one would call a *coup de foudre* between the families. Besides enjoying scientific discussions with Jean-François, Christine, who was a journalist at the local Strasbourg television station, regaled us with the cultural activities to be found in Strasbourg. They urged us to consider spending our next "semisabbatical" in Strasbourg, and Jean-François invited me to work at his institute. I had already tried to return to Europe (see Chapter 13) on a permanent basis but found the administrative burden in France so overwhelming that I had come back to the United States.

Marci, Mischa, and I spent the spring semester of 1992 in Strasbourg and discovered that it is indeed a wonderful place to live, easier in many ways than Paris, and more interesting than might be expected for a city of its size. This was partly due to its history, having been part of both France and Germany at different times, but also because the European Parliament meets there for one week each month. The diplomats tend to work a standard day and often go to the opera, to plays, and particularly to musical events in the evening.

After we had been in Strasbourg for a few months, we—it's perhaps more appropriate to say Marci—thought that we might really decide to move there and began looking for an apartment. I was not enthusiastic, to put it mildly (at that time, I was not ready to think about giving up my Harvard appointment and moving to Europe), but agreed for Marci to go ahead. To my great surprise, Marci found a wonderful apartment with a lovely view that was located close to a park (the *Parc de Contade*) along a small rivulet (Figure 18.1).

The location was within walking distance of the university, which was important to me. It meant that I could walk back and forth, one of the few chances for me to exercise.

Before that initial semester was finished, we had purchased the apartment. We then returned to Boston for the next two years and moved permanently to Strasbourg in 1994. The following year, we undertook to renovate the apartment. Our idea was to open up the interior and create a modern kitchen and inviting open spaces, analogous to those in our Cambridge house. We asked Bill Gaynor, a friend who had been our architect for the Cambridge renovations in 1983, to draw up plans for the Strasbourg apartment. We sent him detailed measurements and photos, and he drew up plans without having actually seen the apartment. The general spirit of the design was a repeat of our Cambridge house, with which we were very happy. Bill managed to retain the feel of an early 1900s apartment with a modern sense of openness. He never saw the apartment before, during, or after the renovations. However, with the help of a local architect and a group of experienced workmen, including the carpenter Jean-Luc Sifferlin who was

Figure 18.1. Swan with cygnet at the edge of the rivulet

essential, the apartment became a wonderful place for us to live. This and our Cambridge house were our homes for the next twenty years (Figures 18.2 a and b).

While living in Europe, we took many trips, often revisiting places that I had seen during my travels in the 1950s and 1960s (see Chapter 17). One place I had not been was Spain, as I never wanted to go while Franco was in power. Mischa spent part of the summer of 1998 in San Sebastian, attending a language program and improving his Spanish. We visited him there and saw what a wonderful city it was (Figure 18.3). Marci and I returned in 2000 when I was invited to give a lecture and again in 2016.

Mischa had attended a bilingual French school in the United States and also attended the local elementary school for first grade in Paris, when I was there on sabbatical. This meant that he had a good grasp of French when we moved to Strasbourg. My two daughters from an earlier marriage were already living independently. My older daughter, Reba, was attending medical school in Jerusalem and my younger daughter, Tammy, had graduated from the Tufts Medical School and started a residency in Iowa.

We lived in Strasbourg until June 1999, when Mischa graduated from the International School (École des Pontonnieres). Actually, Marci and Mischa lived there full-time and I commuted between Harvard and Strasbourg. For several years, I taught half-time at Harvard and arranged my lectures, so I could give a month's

(a)   (b)

Figure 18.2. Strasbourg apartment. (a) Living room. (b) View from dining room window

Figure 18.3. Mischa, Marci, and I in San Sebastian, 1998

worth in about two weeks, which enabled me to spend time with my family in Strasbourg. I formally retired (i.e., stopped teaching lecture courses at Harvard) after three years at half-time, in accord with an agreement I had made with Henry Rosovsky, the dean of the Harvard Faculty of Arts and Sciences at that time.

After Mischa graduated and returned to the United States to go to college at Georgetown University in Washington, DC, Marci and I realized that we could live where we wanted when we wanted. We decided to "commute" following two "guiding principles." The first was that we would stay long enough so we felt we were really living there. This led to our alternating two months in Cambridge with two months in Strasbourg. We accumulated duplicates of most of what we needed, so we did not have to carry things back and forth. The second was that I would accept speaking engagements only on the side of the Atlantic where we were at any given time; not surprisingly I sometimes broke that tenet, though it often provided me with an excuse to decline an invitation. We initiated this schedule following our return from my year as Eastman Professor in Oxford, 1999–2000, and rigorously maintained this two-month rotation cycle for the next fifteen years. In part because Andrew Griffith and Nicolas Winssinger, the colleagues with whom I had mostly interacted, had left Strasbourg, we decided to no longer spend every other two-month period there. This was made official when we sold our apartment in Strasbourg in 2016.[2]

I was first appointed Professeur Associé in the Chemistry Department of the Université de Strasbourg, but when Jean-Marie Lehn founded his Institut de Science et d'Ingenierie Supramoleculaires (ISIS), he invited me to join as a senior scientist, the equivalent of a professor. I was given the title, which I still hold, of Professeur Conventionné. It provides me with all the privileges of a professor in terms of receiving grants, having students and postdocs, and teaching, but is unpaid.[3]

ISIS was a very special place and in many ways unique in France. It was much more like a chemistry department in the United States, which of course Jean-Marie was aware of from his time at Harvard in the 1970s. The most important innovation is that it appointed junior group leaders, the equivalent of assistant professors in the United States, who had temporary appointments for five or so years, but who were completely independent. They were given all the funds they needed to start a group and had much lower teaching loads than those customary in French universities. The idea was that they should have the opportunity to do the best research of which they were capable. At the end of the five-year period, they hopefully would have published enough to find a good position elsewhere. This scheme was successful in most cases.

Unlike my earlier stay in Paris, most of the Harvard students did not come with me, and most of the students I had in Strasbourg were recruited by me when I was in France. I was soon able to create a group of outstanding postdocs, using funds that were provided by ISIS and by grants from the European Union,[4] as well as

---

[2]While in Strasbourg, we had bought shares in a vineyard, Chateau Villars-Fontaines, in an area of Burgundy, called Hautes-Côtes de Nuits. Each year, we received about one hundred bottles of red and white wines at cost (i.e., for the bottles themselves and the work of producing the wine). When we sold the apartment, we shipped the accumulated wine from our cellar to our American address in Massachusetts. After it arrived in the United States and went through customs, we received the attached notice. While it is not clear why we were granted this exception, I presume it did not hurt that I had mentioned on the application that I had recently received the Nobel Prize (Figure 18.4).

[3]The position of Professor Conventioné exists only at the Université de Strasbourg. It was created by Guy Ourisson, when he was president of the then Université Louis Pasteur. In addition to being an outstanding scientist, he had the wonderful quality that if there was a problem, he would find a solution, even if involved doing something that had not previously been done in France.

[4]When I first applied for European Union (EU) grants and students applied for fellowships, my group did well because the criterion was the excellence of the applicant and of the research proposal. However, as time went on, the selection became more political and more weight was given to the requirement that the research be of "immediate use" to the EU. This made the application process so onerous that I stopped applying for EU support. Since 2007 (too late for me), the EU has introduced a merit-based system, similar in spirit to the policies of ISIS. Young researchers within two to seven years of their PhD are

COMMONWEALTH OF MASSACHUSETTS
DEPARTMENT OF THE STATE TREASURER
ALCOHOLIC BEVERAGES CONTROL
COMMISSION
239 Causeway Street, 1st Floor Boston, Massachusetts 02114

DEBORAH B.
GOLDBERG
TREASURER AND RECEIVER
GENERAL

KIM S. GAINSBORO,
ESQ.
CHAIRMAN

NO. 4672

Boston, Mass, February 25, 2016

ACTING UNDER THE PROVISIONS OF SECTION 22A OF CHAPTER 138 OF THE GENERAL LAWS, AS AMENDED, (THE LIQUOR ACT), THE ALCOHOLIC BEVERAGES CONTROL COMMISSION HAS THIS DAY VOTED TO PERMIT:

MARTIN KARPLUS to receive 59.4 gallons of still wine and .514 gallons of spirits to 133 Irving Street, Cambridge, MA 02138 as a gift from 27 Quai Zorn Strasbourg, France. These beverages are for his personal use and not for resale.

ALCOHOLIC BEVERAGES CONTROL COMMISSION

APPROVED
ILM

_____

EXECUTIVE DIRECTOR

PERMIT FEE: $35.75

PERMIT EXPIRES: May 25, 2016

Figure 18.4. Massachusetts special permit

those available from my National Institutes of Health grant at Harvard. Thanks to them, my research continued apace and I was able to explore new areas. Some of the results obtained during this time are summarized in Chapter 23.

---

eligible to apply for starter grants of up to 1.5 million Euros for five years. They must be allowed to work completely independently at the institution of their choice.

Chapter 19

My Life as a Chef

———————— ⚭ ————————

An important part of our life in Europe was the accessibility of good food, not only that in outstanding restaurants but also the fresh produce that varied with the seasons at the local outdoor markets. In the part of Strasbourg where we lived there were no supermarkets. We were lucky to have a small local outdoor market twice a week (on Tuesday and Saturday) and a much larger market about a twenty-minute walk away. Even the local market had three vegetable stands, two meat stalls, and a booth that specialized in poultry products such as maigret de canard, whole chickens, and fresh fois gras. Given that and the local bakery with crusty French bread, we had no need for a supermarket.[1]

Strasbourg was a wonderful place to go out to eat, not only for its starred restaurants, like the Buerehiesel and Au Crocodile, but perhaps even more for its local places with classic Alsatian cuisine. When I ate a dish I liked that I had not had before, I would try to analyze how it might have been made. Sometimes I would ask, though I did not always receive a useful reply. Then I would try to recreate it at home. That such experiences were to lead to my working in famous restaurants was another of the many serendipitous events in my life.

Marci and I were coming to the end of a summer in the Haute Savoie in 1977. She had already flown back to Boston, and I was planning to fly back to the United States when I learned that there was an airplane strike in Paris so that I could not go home immediately. Rather than driving directly to Paris I decided to stop off in Roanne, a small town whose primary interest for tourists was its three-star restaurant, run at that time by Pierre and Jean Troisgros.

For readers not familiar with France and not inclined toward gastronomy, it is worthwhile to mention that the Michelin Red Guide with its listing of restaurants was for decades an essential companion of any traveler in France. The Red Guide defines a three-star restaurant as one with "exceptional cuisine worth a special journey," while a two-star restaurant is one with "excellent cooking, worth a detour." The guide was first distributed for free by the Michelin tire company

———————— ⚭ ————————

[1] Now that farmers' markets with fresh seasonal produce have come to the United States, they make up for much of that aspect we missed from France.

in 1900 as an advertising gimmick to sell tires, when there were only 3,000 cars in France. The star rankings were introduced in 1936 shortly before the Second World War.[2]

In the 1970s, the cooking at Troisgros, like all three-star restaurants, was classic cuisine, i.e., very rich with delicious sauces.[3] The history of Troisgros goes back to 1930 when Marie Troisgros, who was an excellent cook and her husband, Jean-Baptist, who was in charge of the "front of the house" and the wine, started the restaurant. The restaurant received its first star in 1955. Once their sons, Pierre and Jean, began working with their parents, the restaurant received a second star in 1965, and in 1968 it received its third star.

I had a wonderful dinner at Troigrois and, of course, everything was perfect, as it should be in a three-star restaurant. For me, their poached salmon in a rich sorrel cream sauce was something very special. After dinner when the service was slowing down, they invited me and a couple of other straggling customers to sit down with them and share a glass of cognac. I mentioned that I had at least a week with nothing to do because of the airplane strike. After some discussion of my interest in cooking, they invited me to stay for the rest of the week to work in their kitchen. They also recommended an inexpensive inn in the neighborhood where I could stay. I was delighted to accept the invitation. In this first sejour in a three-star kitchen I spent more time watching and asking questions than actually participating. I also met several young chef apprentices, some of whom were children of other three-star chefs. They worked there to get experience in another outstanding establishment before returning to cook at their parents' restaurant. My acquaintance with them helped in some cases to continue my "career" as a chef.

My interest in cooking began at an early age. Already as a young child in Vienna I had liked to spend time in the kitchen. There it was not my mother who usually

---

[2] An interesting sidenote is that when the Normandy landing was being organized in the spring of 1944, there was concern that the Allied troops would get lost or confused in French cities where all street signs had been destroyed by the German troops before leaving. Someone in Washington, DC, had the ingenious idea of reprinting the 1939 Michelin guide (the last prewar edition) with its very detailed city maps and distributing it to the officers of the troops making the D-day landing [https://www.beyond.fr/food/michelin-guide-history.html].

[3] Unfortunately, such cooking now is only rarely awarded three stars, although there is still at least one exception, the Auberge d'Ill, but more about my experience there later in this chapter. My interpretation of this phenomenon is that the food critics, who have to eat in fancy restaurants nearly everyday are always looking for something new. Moreover, there is the current fashion for lighter meals with minimal sauces.

did the cooking, but rather it was Mitzi. Once we had emigrated to the United States, it was my mother who cooked for the family, as we no longer had a maid. In our West Newton home, we had a large kitchen, where my mother was occasionally assisted by my grandmother Mania. Mania had joined us in the United States in 1940, her husband Samuel having died in Vienna in March 1939 from a heart attack. She was able to come to the United States because my parents were able to provide an affidavit for her.[4]

My memories of the kitchen consist of me with my mother and grandmother (Figure 19.1).

My brother displayed no particular interest in cooking and never shared my focus on cooking and food. In the kitchen, Mania and I mainly helped with preparing vegetables and, once in a while, my mother let me help with the actual cooking. She, perhaps from her work as the dietician in the Fango Heilanstalt, had a tradition of cooking simply but with good ingredients, presaging what became the rage in the United States sixty or seventy years later.[5]

I continued to do some cooking at home, in addition to the cooking classes in junior high. My cooking career was interrupted when I went to Harvard, but I took it up again the summer I was at Woods Hole (see Chapter 6), and then when I became a graduate student at Caltech. Alex Rich, whom I had met at Woods Hole, had invited me to share his apartment in Pasadena. What living with Alex meant was that generally we only met at dinner. Alex would usually get up sometime in the afternoon, or even later, and I would cook dinner when I returned from a day in the lab. It was the end of my day and the beginning of the day for Alex, who usually was still in bed when I left in the morning.

The next year, when Gary Felsenfeld, who had been an undergraduate with me at Harvard, came to Caltech as a graduate student, we rented an apartment together and there again, I mainly did the cooking. After about a year in this apartment,

---

[4]Mania spent a third of the year with each of her three daughters who were all living in the eastern part of the United States. My mother was the oldest. Lene, the middle daughter who had been the chief physician at the Fango Heilanstalt, was now a psychiatrist living in New York in a large apartment on Central Park West with her husband Ernst Papanek and their two sons, Gus and George. Claire, a social worker and the youngest (the "baby") sister lived in Utica, New York, with her husband Karl and her two daughters, Lisa and Susan. Alex, the one brother, who was the second oldest of the four children, was living in New Zealand at this time (see Chapter 1).

[5]After my mother had received her a degree from Simmons College, she worked for a number of years as the chief dietician at the Beth Israel Hospital in Boston. When she worked there, the food was good, particularly so for a hospital. (I would go there sometimes for lunch with her.)

Figure 19.1. Mania in the West Newton kitchen

I was invited to join a household where several friends were living, including Alex Rich, Sidney Bernhard, and Roy Glauber. Alex and Roy soon moved out, Alex when he married Jane. Sidney, who later would accompany me on my visit to Lapérouse in Paris (see Chapter 8), was also a very good cook and so the two of us did the cooking; other members of the household, including Walter

Hamilton and Matt Meselson, who came over every so often, took care of the chores, including washing the dishes. At that time I was introduced to horse meat by Seymour Benzer, a phage geneticist in Max Debrück's group. It was a wonderful, inexpensive substitute for beef (though slightly sweet) and I concocted many dishes with it. The law in Los Angeles dictated that horse meat at the supermarket was sold only as meat for dogs. Not deterred by that, horse filet, which cost a fraction of the corresponding cut of beef, turned up in my cooking as horse stroganoff, horse Cordon Bleu, and so on.

As time went on, I gained more experience in the kitchen, so that when other opportunities like that at Troigros came along, I was ready. People often ask me whether there is a connection between my chemistry and my cooking. As far as I can see, there isn't any. In some ways, since I was a theoretical chemist, cooking was the only real chemistry I did.

One of the many attractions of the Haute Savoie, where we were regularly spending summers, was a three-star restaurant, the Auberge de Père Bise in Talloires, which has an ideal setting on the Lac d'Annecy (Figure 19.2).[6]

Figure 19.2. Auberge de Père Bise on Lac d'Annecy

[6]The Lac d'Annecy is a beautiful lake whose water is now considered the purist in France (Figure 19.3). At one time it was very polluted because sewage from the surrounding communities went directly into it. In the 1960s, sewers were installed around the lake. Within a few years, the lake purified itself. It happened so rapidly, in part, because all the water that flows into the lake comes from underground springs.

Figure 19.3. Vista of Lac d'Annecy, 1984

The first time I ate there was in the early 1970s with Bill Reinhardt when we were in Geneva working on a book on many-body theory, which like several other books I have worked on, was never finished.

During our summers at the chalet, Marci and I regularly drove down to Talloires, which had a beautiful public beach. Although our finances were limited, we made a point of having lunch at Père Bise at least once each year, usually to celebrate a special occasion. Like Troigros, Père Bise had begun with the mother, Marie Bise, doing the cooking and her husband François Bise receiving the guests. When we had come to the Haute Savoie, the chef was her grandson, François Bise, who had gained the third star for the restaurant. During the latter years of his life and for some time after his death in 1985, Gilles Furtin was the actual chef. He and Monsieur Michel Marucco, the *maître d'hôtel*, kept the Auberge at the three-star level as long as they were there.

After we had eaten there a number of times, I had gotten to know Gilles because I often chatted with him at the end of the service. One day I asked him whether I could come and spend some time working in the kitchen. Given my "credentials" from the Troisgros "stage," Gilles said yes after checking with Madame Bise, who was not enthusiastic but did not say no. For several summers after that, I would spend one or two weeks in the kitchen, driving down from the chalet in the morning and returning late in the evening. At Père Bise I had graduated to actually working in the kitchen. Often I replaced the person who had a day off. Of course, I had to be shown what I had to do at each "station," but I picked up how each task was done relatively quickly. One day it might be at the *garde manger* preparing vegetables, another day it was helping with making the sauces. Most of the time I did well, but not always. A special dish, one I had eaten on my first visit, was crayfish (*ecrivisse*)[7] from the lake in a rich *beurre blanc* sauce made with butter, eggs, and a little vinegar. It was necessary to be very careful that the sauce "took," as one says in English, namely that it became a delicious thick concoction, without the eggs turning essentially into scrambled eggs. One time when I was stirring the sauce, it did start to scramble and Gilles came over to help. He showed me that by adding a little cold water and stirring rapidly the sauce could be saved. It was a real learning experience. Much later, I saw one of Julia Child's TV shows where in her typical manner, cheerfully claiming "no problem" as disaster loomed, she resorted to the same trick when her *buerre blanc* sauce had turned. It was particularly pleasant working at the Auberge because the atmosphere in the kitchen was more relaxed than in some of the restaurants in Paris where I worked subsequently. This was due primarily to Gilles, though country inns generally tended to have a less stressful atmosphere.

Another cooking experience was at Taillevent in Paris. It had been an elegant three-star restaurant for many years. It was named after a famous French chef who lived in the 14th century and is reputed to have written the first cookbook. Unlike most three-star restaurants, Taillevent was owned and run by Jean-Claude Vrinat, who was not the chef. Vrinat was responsible for making Taillevent a wonderful place both to eat and to work. He watched over everything in the kitchen and the restaurant. In the latter, he made everyone who came to dine there feel that they were his special guests. There were, of course, the regulars or important people that he knew, but customers that he did not know were treated in the same gracious way. After we had eaten there several times, I brought up the question of spending some time working in the kitchen. Vrinat did not immediately say yes and said he would have to check whether there would be a concern about

---

[7]We were very fond of the crayfish and knew that they could be found in the Lac Annecy. One night several friends and I drove down to Talloires with nets and a flashlight. Sure enough, we attracted the crayfish, but before we could actually catch them, the local police arrived, asked what we were doing, and pointed out that catching crayfish without a license was illegal and that there was a hefty fine. We pleaded ignorance and the police told us not to do it again, letting us off without the fine. After that, we only had *ecrivisse* at the Auberge.

insurance; normally all the people who work there are insured for accidents that occur while they are in the kitchen. I was pleasantly surprised when I received a message a couple of weeks later, setting a time to meet and discuss my "*stage.*"

The head chef at Taillevent was Claude Deligne, seconded by Philippe Legendre, who was primarily responsible for me and showed me many little tricks while I worked with him. One evening the kitchen was preparing several simultaneous orders of fillet of veal. It had to be sliced before it was sauced and so I was commandeered to help. I started to cut a fillet, and almost immediately Monsieur Deligne came over to me and guided my arm with the knife while explaining that the veal was so tender that you should not press down on it because that would force the meat juices out; in fact, with a really sharp knife, no force was required (Figure 19.4).

I was very impressed because Deligne had been at the other end of the kitchen and appeared to be just standing around and leaning on a counter. However, he apparently observed what was going on everywhere in the kitchen and made sure that everything was being done correctly. These are lessons that have stayed with me for the rest of my life, not only making sure that my knives are always as sharp as they should be, but also in some sense in watching over the research of my students. Deligne was respected by everyone and, although he was getting up in age, he still commuted on a motorcycle.

An unusual custom initiated by Vrinat was that he encouraged the kitchen staff to invent new dishes. The first step in its acceptance was that the staff, who usually ate before the service started in the evening, were given the chance to taste the dish. If it seemed promising, it would appear as a speciality of the day that was not on the regular menu. Finally, if customers liked it, it would be added to the menu the next time it was updated, with Vrinat making the final decision. Most such proposals were not successful, but some were.

I first visited the Auberge de L'Ill, a three-star restaurant since 1967 in the Alsacian village of Illhaesern, when I was invited to participate in a conference in Basel in 1980. In fact, I had agreed to attend only if we would have lunch at the Auberge, which was about an hour away by car. The organizers took me up on my proposal and I had my first experience of the wonderful food prepared by Paul Haeberlin, who had taken over what was then a local inn from his parents after the Second World War. Being close to the border, German bombardments aimed at a bridge

Figure 19.4. The kitchen at Taillevent

over the Rhine had destroyed the inn; it was rebuilt in 1945. Like Père Bise, the Auberge de L'Ill has a charming location; it is set in a garden along a small river (Figure 19.5).

Once Marci and I were installed in Strasbourg, we visited the Auberge again. While chatting with Jean-François and Christine Lefevre in 1992, I mentioned that I would like to spend time working in the kitchen at the Auberge. Christine,

Figure 19.5. Marci and I at the Auberge de L'Ill in the winter, 2010

who reported on art and culture on the local TV channel, France-3, contacted the Auberge and used her connections to make the arrangements. Thanks to Paul, who was by then in semi-retirement, and his son Marc, who had taken over as chef, the kitchen had a warm and familial atmosphere. What I remember most clearly is Paul gathering up the vegetable cuttings, which had been thrown away because they were not perfect. With some chicken stock, he made a delicious soup that was regularly the lunch entrée for the staff. Having lived through the food shortage during the war, he could not bring himself to let anything go to waste.

One of the specialities was a *terrine de foie gras*. I wanted to learn how to make it, but since it was only made once a week, the timing had not worked out during my *stage* there in 1992. Marc gave me a standing invitation to come on a future Thursday when it was prepared. This finally happened only in January 2015, a few days after the announcement that the Auberge had been awarded three stars for the 45th consecutive year (Figures 19.6 a and b).

While I worked in preparing the terrine, Marc remarked that I was the only Nobel Laureate that he had had as a *commis* in the kitchen (Figure 19.7).

Figure 19.6. *The kitchen at Auberge d'Ill. (a) Making a terrine. (b) With Marc Haeberlin and a commis*

Figure 19.7. *A festive dinner at the Auberge de L'Ill after working in the kitchen*

That the Auberge continues to have three stars is a tribute to the finesse of their dishes. To my knowledge, the Auberge is the only restaurant with three stars that still serves classic, rather than some form of "molecular" cuisine.

Another restaurant experience was with Joël Robuchon. Although there were twenty three-star restaurants in France, Robuchon was considered *the* three-star chef in the 1990s. He had opened a relatively small restaurant called Jamin in the 16th arrondisement of Paris in 1981 and he won his first star the same year; he received the second and third stars in the following two years. This rapid rise was unprecedented in the history of the Michelin Guide. Marci and I dined there several times during visits to Paris. One day after we had eaten there and most of the other customers had left, I asked the *maitre d'hotel* whether it might be possible for me to work in the kitchen for a week or two. I described myself and where I had worked previously. He apparently went to the kitchen and mentioned my request to Robuchon. Robuchon came over to our table and said I should come back when he would have more time. He may have asked Vrinat about me; Taillevent was located only a few blocks away. I returned a few days later to meet with Robuchon prior to the luncheon service. I was asked to sit down and was served a small cup of a delicious cream soup made with cauliflower in chicken stock. Robuchon asked me what I thought of it. I had realized that I was being tested and so had tasted each spoonful carefully. There clearly was something else in the soup with a very delicate flavor, but it was like nothing I had ever had in a French restaurant. Finally, it came to me that it had to be made with sea urchins, which I had eaten previously in Japan and also in Marseille, where women sold them freshly caught from small carts in the harbor. I hesitantly described my conclusion, and apparently he was sufficiently impressed that he said I could come and work in the kitchen at his restaurant.[8]

By the time we had worked out a date (it was in 1995), his restaurant had moved to an elegant townhouse near the Arc de Triomphe and we were living in Strasbourg. Working in each three-star restaurant was a different experience. Robuchon was more of a perfectionist than any of the other chefs with whom I worked (Figure 19.9).

Whenever something new was being added to the menu, he wrote a detailed description of every step in the recipe and made sure that the chefs followed them. One of the classics for which he was famous was "simple" mashed potatoes. They had to be made with a special kind of potato, called "*ratte*," and were whipped with enormous amounts of butter. Each person in the kitchen would add a dollop

---

[8]Many years later when a symposium to celebrate my 85th birthday and the Nobel Prize was held in San Francisco in October 2014 (Figure 19.8), we stayed at David Chandler's house and we had a "pre-dinner" there, which I had arranged for speakers. David and his wife Elaine recommended a caterer that they had used for her 65th birthday celebration. Of course, I wanted to make sure that the food would be up to my standards so the caterer and I corresponded over the details of the menu. I proposed that we should have the cauliflower/sea urchin soup that I remembered from my Robuchon days. I looked on the Web and discovered that San Francisco is a place where one can purchase fresh sea urchins, in part because there is a large Japanese population. I sent the recipe to the caterer, who agreed to make the soup. In the afternoon before the dinner, the chef and his *equipe* came and were working in the kitchen. I went in every so often to taste the dishes as they were getting them ready. They had been preparing the soup when I realized that they were planning to use the sea urchins only as a garnish on top. I said that the whole point was that the sea urchins should be mixed into the soup to give it the very special delicate taste that I remembered. The resulting soup was a great success, as far as I was concerned, though most people probably did not know what the "secret ingredient" was.

Figure 19.8. Karplusians at the symposium

Figure 19.9. Robuchon's kitchen

of butter and whip the concoction when he (there was no *she* in the kitchen at that time) walked by. When I was there it happened to be black Perigord truffle season and many of the dishes were more than amply garnished with them. The truffles were cut into very fine slices (*à la julienne*). Because this took a lot of time, several people were doing the chopping and I went over to help. Robuchon apparently watched me for a while (I did not know that he was doing so) and then came over and politely told me to do something else—it was clear that my technique was not up to his standards, particularly because the truffles were a very expensive item. In fact, the restaurant lost money on the truffle garnish, but Robuchon wanted enough on each dish, such as his cabbage balls stuffed with shrimp (this sounds simple, but they were delicious, and unlike the cauliflower/sea urchin soup, contained ingredients that I never learned), that the truffles contributed significantly to the taste. In some other restaurants, only enough would be used for the appearance of the dish.

As my *stage* was nearing the end, it had been so special that I thought it would be nice to finish by having lunch there with Marci. I waited to request a reservation until a few days before, so not surprisingly when I asked the *maitre d'*, he said he was sorry but the restaurant was fully booked. However, he came back a little while later and said he had managed to find a table for us; presumably Robuchon had intervened. Marci had arrived from Strasbourg by train that morning, and we were ushered to a well-situated table with a bottle of champagne in a cooler waiting for us. Marci's menu did not have any prices. This was not, in itself, unusual in elegant restaurants. However, mine did not list any prices either. I pointed this out to Marci and asked her what I should do. She said, "Well, it's pretty clear to me what is going on. Robuchon is inviting us." It was a wonderful and gracious luncheon, at the end of which the *maitre d'* escorted us to Robuchon's private sitting room where we had coffee and cognac with him and chatted for a while. After we returned to Strasbourg, I sent him a special bottle of *eau de vie de poire* (pear brandy) for which Alsace was famous. He in turn sent me a letter thanking me (Figure 19.10). Robuchon was not only a great chef but also a real gentleman!

In 2000, while in San Sebastian to give a lecture to the chemistry faculty of the University of the Basque Country, I arranged to have lunch at Restaurant Arzak, the oldest three-star restaurant in Spain. It served updated Basque cuisine; their signature dish was the locally caught hake in a green sauce. Thanks to an

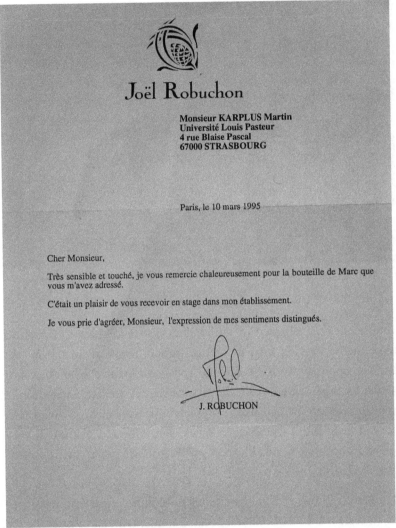

Figure 19.10. Thank you letter from Robuchon

introduction to Juan Mari Arzak, the chef/owner, by Jesus Ugalde, a professor at the university, I was accepted to work in the kitchen and returned to San Sebastian for a *stage* in December 2000 (Figure 19.11).

Juan Mari and his daughter Elena together provided a very pleasant, though at times chaotic atmosphere. When I was awarded the Nobel Prize in 2013, I received an email from Juan Mari and Elena, congratulating me and inviting me the next time I was in San Sebastian to come to the restaurant "in the front

Figure 19.11. In the Restaurant Arzak kitchen with Juan Mari, his wife, and daughter Elena

of the house" as they phrased it, rather than in the kitchen. The invitation to participate in the Passion for Knowledge in San Sebastian in the fall of 2016 provided the opportunity for me to do so. Accompanied by Xabier Lopez, a former postdoctoral student from the Basque region, we had a wonderful lunch at Arzak. The restaurant had been completely transformed. It now included a *laboratory*, *à la* El Bulli, where new dishes were invented. Although the lunch was special, I hope that Arzak, like the Auberge de L'Ill, will return some of their classics to the menu.

In September 2002, I worked in the kitchen of El Bulli. It was my final experience working in restaurants. El Bulli was located in a small village near the town of Roses in Catalonia. It is the restaurant where Ferran Adria made his name and originated "molecular cuisine." I had been invited to give a lecture in Roses and decided to try to obtain an invitation for a *stage* at El Bulli. My son Mischa helped me write a letter in Spanish in which I described my pedigree with emphasis on my being a chemist. I knew that Adria employed many chemical techniques (e.g., he lyophilized *fois gras* making a powder out of it, which he then sprinkled on various dishes to add flavor). As a result, Adria invited me to spend two weeks in the kitchen and also to have a meal in the restaurant. That was unusual because the restaurant was open only during the summer and normally reservations had to be made two years in advance. The meal itself consisted of nearly forty courses, each of which might be just a bite or two (Figure 19.12).

Working at El Bulli was also special, in that the atmosphere was very welcoming because almost everyone in the staff of over forty worked there for the experience without pay. So it was easy to learn—you had only to ask to help out on a dish to immediately be shown how to prepare it. Of course, the assumption was that you

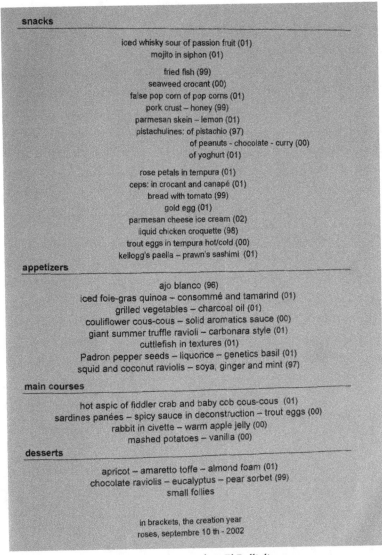

snacks

iced whisky sour of passion fruit (01)
mojito in siphon (01)

fried fish (99)
seaweed crocant (00)
false pop corn of pop corns (01)
pork crust – honey (99)
parmesan skein – lemon (01)
pistachulines: of pistachio (97)
of peanuts - chocolate - curry (00)
of yoghurt (01)

rose petals in tempura (01)
ceps: in crocant and canapé (01)
bread with tomato (99)
gold egg (01)
parmesan cheese ice cream (02)
liquid chicken croquette (98)
trout eggs in tempura hot/cold (00)
kellogg's paella – prawn's sashimi (01)

appetizers

ajo blanco (96)
iced foie-gras quinoa – consommé and tamarind (01)
grilled vegetables – charcoal oil (01)
coulliflower cous-cous – solid aromatics sauce (00)
giant summer truffle ravioli – carbonara style (01)
cuttlefish in textures (01)
Padron pepper seeds – liquorice – genetics basil (01)
squid and coconut raviolis – soya, ginger and mint (97)

main courses

hot aspic of fiddler crab and baby cob cous-cous (01)
sardines panées – spicy sauce in deconstruction – trout eggs (00)
rabbit in civette – warm apple jelly (00)
mashed potatoes – vanilla (00)

desserts

apricot – amaretto toffe – almond foam (01)
chocolate raviolis – eucalyptus – pear sorbet (99)
small follies

in brackets, the creation year
roses, septembre 10 th - 2002

Figure 19.12. Menu of my El Bulli dinner

were already an experienced chef, interested in learning something new. In most cases, I knew enough to be accepted.

Since 2005, I have restricted my cooking to my home kitchen, where Mischa helps me when he is around, and Marci invariably cleans up. Perhaps the most important thing I learned from my restaurant experience is not how to make any specific dish, but rather how to pace the preparation of the various dishes so everything is ready at the right time. Often as I walk home from my "lab" at Harvard, which takes about ten minutes, I decide what I will make that evening, taking account of any leftovers and sauces that we have in the refrigerator. I find it relaxing after an intense day at the office to prepare the dinner, which usually, but not always, is a success. These days, we tend to celebrate family birthdays at home, cooking a special menu. We have our favorites, for which we are now able to find the ingredients locally in Cambridge. Examples include *fois gras poêlée* with pears or apples, seared duckbreast, and New England bay scallops and lobsters. One special dish, which I first prepared in 1978, is a scallop mousse with Duglere sauce, based on a recipe by Craige Claiborne in *The New York Times*. What made their version easy is that the mousse could be prepared by putting everything together in a food processor, instead of using sieves and iced ingredients.

My experiences with cooking would not be complete without mentioning Paul April, who I first met on my 60th birthday. Marci, having been incapacited by foot surgery, phoned our neighbor Julia Child and asked her to recommend a caterer. Marci contacted him and explained that she would like their main chef to cook a dinner for a few people. When I came home from work the day of my birthday, I saw a young man working in "my" kitchen. I was none too pleased to see this and expressed my annoyance: "Who is in my kitchen?" Paul introduced himself and proceeded to explain what he was doing. Before I even had a chance to taste his cooking, he had won me over. Now over 25 years later, we stay in touch, occasionally preparing an extravagant meal for our families.

Chapter 20

Announcement of the Nobel Prize

———— ⚜ ————

lthough the Nobel Prize committees make very clear that people who nominate someone for a Nobel Prize must not contact the individual, in my case at least I was well aware that I had been nominated various times by numerous individuals for over twenty years. In the 1990s, while living in Strasbourg, I would be at my computer at 11am the day of the chemistry prize announcement, working as usual but also waiting to find out who had been awarded the prize that year. After some years, I stopped this practice, having concluded that I would never receive the prize. The work on which I felt it should have been based was done in the mid-1970s and so it seemed that my chance had passed.

Once the Nobel Prize was announced on October 9, 2013, my life changed. How it would change, fortunately, would in part at least be up to me. But more of this later. That morning we were in Cambridge, and while I was still on the phone with Sweden, the first sign that the outside world had also been informed was the arrival of Stephanie Mitchell at our front door well before the sun was up. As a photographer for the *Harvard Gazette*, she was aware that it was not unlikely that a Harvard professor could be a recipient, and she was diligently listening to the Nobel announcements on the morning (5am) news. Since she happened to live a few blocks from us, she wasted no time. My son, Mischa, answered the doorbell and invited Stephanie in. Marci had just stepped out of the shower and, hearing voices in the entryway, went to see to whom Mischa was talking. Being introduced to Stephanie as a reporter, Marci gasped, "You are media." She then politely asked her to wait out on the porch until we were presentable. Marci called out to me to get off the phone and get dressed. About twenty minutes later, we invited Stephanie back in to join us for breakfast. Her presence resulted in a detailed report of what a Nobel Prize winner, or at least one Nobel Prize winner, eats for breakfast on the day he receives the news (Figure 20.1).

We had our typical breakfast of bacon, cheese, toast, and black coffee, all of which I usually prepared, but that morning it was Marci who made breakfast. One photograph (Figure 20.2) by Stefanie was circulated worldwide (Figure 20.3).

Figure 20.1. Having breakfast, with Mischa looking on
[Reproduced with permission from Stephanie Mitchell]

Figure 20.2. Sitting in our living room

I received an incredible number of phone calls that morning, including one from the Austrian Radio (ORF), with whom I had my first interview in German (Figure 20.4).[1]

---

[1] The interviewer asked me what my feelings were about Austria. I responded by saying that Austria had expressed no interest in me or my science in the 75 years since I had left. Also, I mentioned, as already described in Chapter 17, my experience in Vienna when I was there for a meeting in November 1988, the 60th anniversary of Kristallnacht, the night in Germany and Austria when hoodlums went around and shattered the windows of Jewish-owned stores.

Figure 20.3. Marci with Bib, Mischa, and me

Figure 20.4. The telephone interview

Although it is my native language, I have rarely used it. In a sense I have the vocabulary of an eight year old, but an essentially perfect accent. This is in contrast to my French, in which I am fluent but do have an accent. (I am pleased when people ask whether I am from Belgium.)

A celebration later that morning was organized by the Chemistry and Chemical Biology Department (CCB) and it was held in the departmental library (Figure 20.5), a beautifully paneled room that still retains much of its old charm. Champagne was served and several congratulatory speeches were made by my colleagues (Figure 20.6).[2]

Midway through the festivities, the Harvard University Provost came in with a bouquet of roses (Figure 20.7a). He had no idea who I was when he came into the room. After we were introduced, he made congratulatory remarks including a message from President Faust. I commented that I had recently been informed by the chairman of the department that there was no guarantee that I could keep the already reduced space presently occupied by my group since all space technically "belonged to the University as a whole." I asked the Provost whether he could give me a guarantee (Figure 20.7b).

He reassured me with an "of course." I believed that to be helpful, but knew from past experience that even an agreement in writing from the Provost or Dean of the Faculty of Arts and Sciences was not "iron-clad."

Figure 20.5. Interview in the CCB Library

[2]I remember one, in particular, by Stuart Schreiber, who said how important it was to him that I had encouraged him to focus his research on biologically related problems as he wished, in contrast to what the department had hired him for. He was brought in from Yale to continue the department's strength in synthetic organic chemistry, a legacy of R. B. Woodward. Of course, given Schreiber's scientific work, there no longer are any complaints.

Figure 20.6. George Whitesides giving a congratulatory speech

(a)   (b)

Figure 20.7. Visit from Provost Garber. (a) The Provost with roses. (b) Me with the Provost

During the gathering, E. J. Corey, Nobel Laureate in Chemistry 1990, remarked how lucky I was to have received the Nobel Prize when I was 83 years old, so that I had had about 25 more years than he had to peacefully continue in research. Corey invited me to stop by his office for advice on the upcoming ceremonies in Sweden. I wondered what Corey might say that would be useful. It turned out his advice did much to make the Nobel Prize event in Sweden a pleasant experience and proved invaluable in navigating the pre- and post-celebration world. The essence was that I had received the Nobel Prize, and so it was up to me to decide what I wanted to do or not do. Illustrative of this was his point concerning the requirement that one wear patent leather shoes for the award ceremony. He had worn nice black shoes and no one had complained. Finding shoes to fit an octogenarian's flat feet is not easy! Thus, at the award ceremony, neither I nor my fellow octogenarians, Engelbert and Higgs, wore patent leather shoes. Also, Corey warned me that I would be told that there were certain events outside the official ceremony in which I must participate, but again it was up to me to decide in which I would actually do so.

Chapter 21

After the Announcement

———— ⟨ ❦ ⟩ ————

Shortly after the announcement of the Nobel Prize, we received an invitation to come to Washington and meet President Obama at the White House. Apparently it was a long-standing tradition that the President invites the American Laureates; this tradition was discontinued by President Trump. I was thrilled by the idea of meeting President Obama but wanted to know more about what this meeting was likely to entail. I contacted John Holdren, the science advisor to President Obama, for information about such a meeting and said that if it just a photo-op, I was not interested in going. He understood my feeling, but diplomatically said that when you are invited by the President, normally you refuse for only two reasons: (1) you are dead, or someone in your family has recently died, or (2) you are too ill to come. He then commented on what the meeting might involve and encouraged me to come. He said that the President would decide how much time he would spend with me and the other American Nobel Laureates. After this conversation, Marci and I decided that we would go to Washington, and it was indeed an honor to meet President Obama.

Given our aversion to formal arrangements, especially first thing in the morning, Marci and I had skipped a breakfast gathering, choosing instead to walk directly to the White House where we would meet up with the rest of the Nobel Laureate group. We arrived early and were rapidly checked through security and then admitted to a waiting room in the West Wing of the White House, just outside the Oval Office. As we sat there, various officials, including Senator McCain and Secretary of State Kerry, passed by us and entered a meeting room next to the Oval Office. We could hear animated, even if indistinct, discussions going on for the next 45 minutes. Finally, we saw the same group file out. Subsequently we learned that this had been an emergency meeting to discuss the report of Secretary Kerry on the progress of the nuclear deal with Iran. This was a reminder that amidst what to us was an important event, namely an audience with the President, far more urgent and complex decisions for the future of the country and the world were being debated on that November 19, 2013, the date of our visit.[1]

———— ❦ ————

[1] The negotiations with Iran were at a critical stage. On September 28, 2013, President Obama held the first direct talks with Hassan Rouhani, the Prime Minister of Iran, and on November 24, 2013, the five permanent members of the UN Security

Once the President was ready to receive us, the Nobel Laureates were lined up outside the Oval Office according to the Nobel protocol. Since no American had received the Physics Prize that year, the Chemistry Prize winners were first in line. Further, for each prize, the winners were lined up alphabetically. Thus, Marci and I were to be the first to enter the Oval Office, and of course it was "ladies first." As the door opened and Marci entered, President Obama was standing in the doorway and leaned over to shake her hand, saying, "Hi Marci, how are you?" (Figure 21.1).

The same warm welcome was given to each person until we all had entered the Oval Office and were lined up, once more in "canonical" order (Figures 21.2 and 21.3).

When I was standing next to President Obama, I took the opportunity to say to him, "You are the second president that I've met. The first one was President Truman who I met when I was in Washington as a Westinghouse Scholarship winner in 1947." Further, I commented that Truman was the first president to

Figure 21.1. President Obama greeting Marci

Council (USA, UK, France, Russia, and China) plus Germany reached an interim agreement limiting Iran's nuclear program in return for partially lifting the sanctions against Iran. It took until July 2015, to obtain the final agreement. Although inspections showed that Iran was complying with the agreement, President Trump announced that the United States was withdrawing from the Iran agreement on May 9, 2018.

Figure 21.2. Nobel Laureates in Oval Office

Figure 21.3. Marci and I with President Obama

try to get a bill passed to introduce universal health insurance. President Obama paused and was clearly thinking. He then said, "No, it was President Theodore Roosevelt who first tried to do this." Almost immediately he corrected himself, "I was wrong. Roosevelt did indeed propose universal health insurance but that was when he was a candidate and lost." I responded, "You are the president that actually made it happen." Although this was a small incident, it showed his awareness of history and his focus on us, in spite of all the important things that were going on, as mentioned above. It made a great impression on me.

During the meeting, President Obama emphasized basic research and the critical role that federal funding played in supporting Nobel-worthy discoveries. He noted that some of this year's American Laureates were immigrants, emphasizing the importance of making America attractive to immigrants. As I write this in 2019, I am struck by the timeliness of those remarks, given the present situation in the United States.

I had brought the catalogue from my Bibliothèque Nationale de France photographic exhibition (see Chapter 17), in which I had written a dedication to President and Michelle Obama. I gave it to him and he leafed through it briefly before handing it to an assistant, as I presume happens to most such gifts.

Another event during our trip to Washington was a "public interview" with all of the Nobel Laureates. It took place at the Swedish embassy, which has an idyllic setting overlooking the Potomac River. Lined up again in the "canonical" order, we sat in front of an audience composed of reporters and other guests who had been invited to participate. After we chemists and the physiologists each had said something to explain the Nobel award in "layman's language," it was the turn of the economists whose award citation read "for their empirical analysis of asset prizes." I turned to them and remarked, "The stock market rises one day by two hundred points and the next day it drops by two hundred points. It seems to me if you really understood the stock market, you would be able to predict this ahead of time." Eugene Fama's immediate reply was essentially that I did not understand his analysis of the stock market. This may be true, though I suppose what he meant is that he is concerned with long range trends, rather than short term fluctuations. In fact, the three winners (Eugene Fama, Larse Peter Hansen, and Robert J. Schiller) have very different views of how to "understand" the behavior of the stock market, in sharp contrast to the complementary views of the three chemistry and the three physiology/medicine prize winners.

Today one wonders why Alfred Nobel did not include a biology prize. Granted, that when Nobel made his will in November 1895, biology was less important than it is today. The highly significant discoveries in biology are now rewarded either by the Physiology/Medicine Prize or the Chemistry Prize. Objections have been voiced by chemists that some of these Nobel Prizes are for achievements that are important only because of their biological impact, rather than chemistry. Perhaps like the Economics Prize, there will one day be a Biology Prize "in memory of Alfred Nobel."

Alfred Nobel also wrote in his will that the Nobel Prizes should be awarded to "those who, during the preceding year, shall have conferred the greatest benefit to mankind." The Royal Swedish Academy of Sciences, which awards the physics and chemistry prizes, certainly uses a very broad interpretation of this clause. As an example, the Chemistry Prize, which was awarded to the three of us "for the development of multiscale models for complex chemical systems," is based on research that was done in the 1970s, nearly 40 years earlier.[2]

A gala dinner held at the Swedish ambassador's residence was preceded by a reception, which had so many people milling around that there was hardly room to move. At the end of the reception, we were shown to our assigned places at the dinner tables. I was seated between the recently named Federal Reserve chairwoman, Janet L. Yellen, and Supreme Court Justice Ruth Bader Ginsberg. I could not have asked for better dinner companions. Since I was a guest of honor, the conversation turned to my life. After some discussion, I mentioned the 2006 "Spinach" article [**Karplus, 2006**] and later I sent copies to both Janet Yellen and Justice Ginsberg. Justice Ginsberg wrote me a note saying she was looking forward to reading it, but that she would have to wait until after the end of the term because she was too busy with the cases coming up. Janet Yellen wrote that she had enjoyed reading the article and that one of her best friends was the wife of Leonard Nash, the chemistry teacher at Harvard who played such an important role in my undergraduate education (see Chapter 6). This reminded me of the small-world experiments [**Milgram, 1967**] that suggest that any two people know each other through a chain of no more than three or four people. Interestingly, the small-world type of connectivity has also arisen in some of my work on protein folding [**Vendruscolo et al., 2001**].[3]

---

[2]The Physics Prize that same year was awarded to François Englert and Peter Higgs for their theoretical work proposing that there exists a particle, now named the Higgs Boson or the God particle, which explains why other particles have mass. Their theoretical work was done in 1964, but the prize had to wait until the theory was confirmed at the CERN laboratory in 2012. In that sense, theory in physics is not all that different from theory in chemistry, as I described in Chapter 12. That the prize was divided between Englert and Higgs and not into thirds, including CERN, was a poignant recognition of the fact that Robert Brout, who collaborated on the theory had died in 2011. Posthumous Nobel Prizes are forbidden.

[3]My granddaughter Rachel learned from a genealogical site that she was a tenth cousin twice removed from Anne Frank. This does not satisfy the small world principle, but is still interesting.

Soon after the Nobel Prize announcement, I received several invitations, including those from Strasbourg, one from Israel, and one from Vienna. The first event took place in Strasbourg on November 13, 2013, and was a small but warm celebration by the Université de Strasbourg in my honor. The president of the university, Alan Beretz, and Jean-Marie Lehn, my long-time colleague and Nobel Laureate in Chemistry, 1987, gave brief speeches followed in true French style by a delicious "apertif" luncheon (Figures 21.4 and 21.5).

About a year later, in September, 2014, I was inducted into the Legion d'honneur as a "Commandeur" (Figure 21.6). I was particularly pleased that Jean-Marie Lehn made the award presentation, as the delegate of the President of France, Francois Hollande. Jean-Marie had been named "Grand Officier de la Legion d'honneur" at the same time.

President Hollande actually had visited Strasbourg in January 2014 to discuss the future appropriations for the support of science and I had the opportunity to chat with him (Figure 21.7). In a way, our interaction mirrored that with President Obama, including my presenting him with the BnF catalogue.

In January 2015, Marco Cecchini, a former postdoc, who was on the ISIS faculty at that time, arranged a symposium in my honor (Figure 21.8).

(a)   (b)

Figure 21.4. Strasbourg reception. (a) Alan Beretz, Jules Hoffman (Nobel Prize in Physiology and Medicine, 2011), and Jean-Marie Lehn with me. (b) Strasbourg Karplusians who attended the reception: Roland Stote, Marco Cecchini, Annick Dejaegere, Michael Schaefer, Fabrice Leclerc, Tom Simonson, Emmanuele Paci, and Stefan Boresch

Figure 21.5. *The reception in the ISIS atrium with my photographs on the wall*

Figure 21.6. *Wais Hosseini and Louisa de Cola, named Chevalier, Jean-Marie Lehn, named Grand Officier, with me*

Figure 21.7. President Hollande, with Jean-Marie Lehn and me

A number of colleagues gave presentations. Among them was Jean-Paul Malrieu, a well-known quantum chemist whom I had first met at the Institut de Biologie Physico-Chimique, (informally known as the Pullman Institute) in Paris during a sabbatical in 1972. As part of my stay there I learned French by requiring anyone who wanted to discuss science with me to speak French; soon my spoken French was better than their English. Another lecturer at the Strasbourg symposium was Gerhard Hummer. When he was interviewed after the lectures, he remarked that there should be a second Nobel Prize for molecular dynamics, echoing my views about the Nobel citation. Marco had asked me to give a closing talk at the symposium. Initially I wondered what could be appropriate for me to say, but I came up with the idea of using the last scientific slide of my Nobel Lecture (see Appendix 3) as a foil. The title is "What Does the Future Hold?" and, like the "Marsupial Lecture" (more on this later), I have elaborated its content over time and used it on several occasions.

From Vienna I received communications asking about my possible participation in events organized to celebrate the 650th anniversary of the founding of the University of Vienna in 2015. Marci and I made reservations to stay in Vienna for three weeks at the Hotel de France, a very comfortable hotel conveniently located within walking distance of the main building of the university. A primary

Figure 21.8. Poster for minisymposium

focus of our choice was that the hotel allowed dogs, and our cockapoo, Bib, was to accompany us. At the last minute, Marci had to delay coming as she had come down with a severe case of the flu. Thus, she and Bib joined me in Vienna a week later and missed the first event on May 8, which was my induction into the Austrian Academy of Science. Anton Zeilinger,[4] the president of the

---

[4]I originally met Anton Zeilinger in the spring of 2013, when he was inducted into the National Academy of Sciences. I had attended the induction ceremony because my long-time friend and colleague Christopher Dobson was elected as an honorary associate the same year. As a new member, one is asked to sign the signature book, a custom adopted from the Royal Society

Academy, who had graciously flown to Stockholm to participate in my visit to the Austrian consulate there, wrote to me asking whether I would accept the honorary membership, realizing that with my forced escape from Nazi Austria in 1938, I might have mixed feelings. As with all these Austrian honors I was somewhat ambivalent. There was also the fact that in the intervening 75 years no Austrian institution—governmental or academic—had shown any interest in my work, as well as Austria's continued demonstration of anti-Semitism. However, the sincerity in which these honors were offered, in particular by Minister Joseph Ostermayer and Professor Anton Zeilinger, convinced me that bearing witness to what had happened could be a positive contribution. By so doing, it would hopefully make it less likely that the next generation would commit similar atrocities.

During our stay, I was awarded an honorary doctorate by the University of Vienna. In addition, there were two events that were especially important to me. The first was an exhibition of my BnF photographs at the university for two months. This was arranged thanks to Minister Ostermayer, whom I had first met while he was attending the Herbert Kelman Conference on "The Transformation of Intractable Conflicts" in March 2014 (see Chapter 3). He opened the exhibition in Vienna with a very warm address, expressing his desire to help make my visit to Vienna a success (Figure 21.9).

The second was being named an honorary citizen of Vienna by Vienna's Mayor Häupl. In his presentation speech, Mayor Häupl said that what had been done could not be undone, but that Austria must make sure that it would never happen again. When I gave my acceptance speech, I remarked that in the US being named an honorary citizen corresponds to being given the key to the city. Further, that one wish in connection with this was that our dog Bib could enjoy the city with us, namely by being allowed into the museums. That evening, an envelope was delivered to our hotel room with a 4 × 5 inch certificate that had an image of Bib, duly signed by the mayor stating that our dog was to be allowed into museums (Figure 21.10).

---

of London. When I was elected to the National Academy in 1967 at the height of the antiwar protests, I had put forward a motion condemning the United States participation in the Vietnam War. When this motion was rejected, I was so upset that I refused to sign the membership book. At the 2013 meeting, I mentioned this and the Academy secretary found that indeed my name was listed without my signature. Just over 45 years after my initial induction into the Academy, I did sign.

Figure 21.9. Anton Zeilinger, Joseph Ostermayer, and Rector Heinz Engl with me at the exhibition

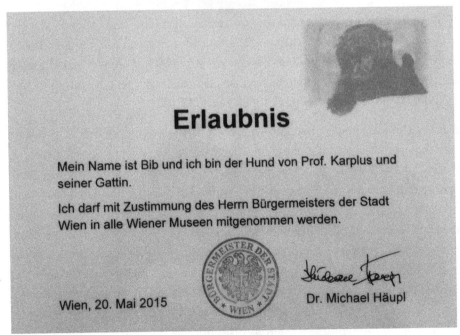

Erlaubnis

Mein Name ist Bib und ich bin der Hund von Prof. Karplus und seiner Gattin.

Ich darf mit Zustimmung des Herrn Bürgermeisters der Stadt Wien in alle Wiener Museen mitgenommen werden.

Wien, 20. Mai 2015

Dr. Michael Häupl

Figure 21.10. Bib's certificate from Mayor Häupl

After the formal speeches and presentation, we were shown into the anteroom of the grand city hall where marble plaques in the wall are inscribed with each of the names of those designated an honorary citizen of Vienna; the plaques go back to 1797 and include Joseph Hayden (1804) and Oskar Kokoschka (1961), among others (Figures 21.11 a and b). The permanency of this inscription had a powerful effect on me.

Figure 21.11. (a) Mayor Michael Häupl and Hannes Androsch in front of the plaque (credit. C. Jobst). (b) Andreas Mailch-Pokorny, Mayor Häupl with Marci, Bib and me in front of the plaque (credit. C. Jobst)

While in Vienna, we also visited the Jewish Museum (see Chapter 3). Under the direction of Dr. Danielle Spera, it had been transformed and had begun to hold a wide ranging series of exhibitions. When we were there, there was an exhibition describing the creation of the Ringstrasse (Figure 21.12), which also addressed my family and the seizure of our property in 1938.

In Israel I was invited by Bar Ilan University to receive an honorary doctorate, with the ceremony to take place on May 20, 2014, following a public lecture, which I planned to be the "Marsupial Lecture." Given my views about Israel's treatment of Palestinians, I was not sure I wanted to accept. I decided that I would get names of Palestinian universities with chemistry departments and request that scientists from each of them be invited to attend my lecture and the ceremony. To my pleasant surprise, the president of Bar Ilan, David Herschkowitz, agreed to do so and I accepted the invitation to be awarded the honorary doctorate. However, as it turned out, none of the Palestinian scientists agreed to attend. Some offered no explanation, and those that did thanked me for the invitation and cited the humiliating and onerous procedure involved in being permitted to cross the border into Israel (Figure 21.13).

After some thought, I concluded that I could make a positive gesture by presenting the "Marsupial Lecture" at Al-Quds, a Palestinian university, the day before the Israeli ceremony (Figure 21.14). One of the editors of *Haaretz*, a liberal Israeli newspaper, invited me to write an Op-ed article describing this series of events; it is reproduced in Appendix 4.

(a)   (b)

Figure 21.12. In the Jewish Museum. (a) Standing in front of the entrance to the exhibition about the Ringstrasse. (b) Danielle Spera with Marci and me (photographs by Sonja Backmayer)

Figure 21.13. Checkpoint for Palestinians to enter Israel

Figure 21.14. Meeting with faculty members at Al-Quds University

At another serendipitous event in my life, I met Gudrun Kramer at the 2014 Herbert Kelman Conference on the Transformation of Intractable Conflicts. She was the Director of the Austrian Study Center for Peace and Conflict and Program Manager for the Regional Social and Cultural Fund for Palestinian Refugees and Gaza Population. When she learned that I was going to Israel to receive an honorary degree and we had discussed my sentiments, she invited me to visit the West Bank with her. She could travel there because she has Jordanian diplomatic license plates. I considered this a wonderful opportunity and we settled on May 18, after I would receive the honorary degree from Bar Ilan University on May 14, 2014. Among other areas we visited Hebron, a divided city with a few hundred Jewish settlers protected by a large contingent of Israeli soldiers; the Palestinian population is about 200,000. (Figures 21.15 through 21.17).

Figure 21.15. Market in front of Damascus Gate to Old City of Jerusalem

Figure 21.16. Graffiti on Palestinian side of Separation Wall

Figure 21.17. Graffiti on Israeli side of Separation Wall

Chapter 22

The Nobel Prize Event

ome days after the October 9, 2013, announcement, we were contacted by the Nobel office in Sweden concerning the arrangements for our trip to Stockholm and our stay there during the Nobel Prize ceremonies in December. We were assigned Eva Nilson Mansfeld from the Swedish Foreign Ministry, who would be responsible for us during the Nobel week. This turned out to mean two things: she would help us by answering all our questions, and equally important, she would make sure that we did what was expected of us.

One aspect of the pending Nobel Prize ceremony immediately became a top priority in my mind. Each Nobel Prize winner is supposed to present a lecture approximately thirty minutes in length describing the history of the research leading up to the award. In thinking about the lecture, I contacted Professor Gunnar Karlström, chairman of the five-member committee in charge of selecting the persons to receive the Chemistry Prize (although the "official" decision is made by a vote of the entire Swedish Academy). I talked with him about the background information that had been passed out to the press and the Nobel Prize citation itself, which read "for the development of multiscale models for complex chemical systems" (Figure 22.1).

He immediately recognized that what I thought was important in my work was the development of molecular dynamics simulations for the studies of biomolecules, but he reiterated that the prize was *not* awarded for that. Moreover, he stated that the committee wanted to narrow the area for the Chemistry Prize to three people.[1] That was the reason for choosing "multiscale modeling for complex chemical systems" to which Levitt, Warshel, and I had made significant contributions in the 1970s. (This aspect of my work is discussed in more detail in Chapter 12.)

As I mention in the printed version of my Nobel Lecture (see Appendix 3), Albert Einstein was upset that he was given the Nobel Prize in Physics (1921) for his theory of the photoelectric effect, rather than for his general theory of relativity, which had already been verified by experiment and which he, as well as

---

[1] The Nobel Prizes are limited to at most three individuals, according to the will of Alfred Nobel. In recent years, the fraction of the prizes awarded to three people in the sciences, though not in literature, has increased.

Figure 22.1. Poster for Nobel Awards

other physicists, regarded as his most important achievement. He was sufficiently annoyed that not only did he give his Nobel lecture in Gothenburg, rather than Stockholm, but he devoted the lecture to the discovery of general relativity [**Friedman, 2001**]. I too felt that my most important contribution to science had been specifically ignored by the committee. Inspired by Einstein, I chose to focus my lecture on the development of the molecular dynamics simulation method.

Returning now to preparations for going to Stockholm, I did decide to conform to the requirement to wear a tuxedo for the Nobel dinner and the prize presentation ceremony, and to wear a dark suit with a tie to deliver my Nobel Lecture. For most other events during the week in Stockholm, a "business suit" was listed as the required attire. Such "requirements" were explicitly spelled out in our information pamphlet. My idea of a "business suit" is that I wear dark-gray wool trousers, a pin-striped shirt, and a V-neck sweater rather than an actual suit.[2] One such gathering was the "Informal Get-Together" on the day after we arrived, where the Laureates were supposed to wear a "business suit," while guests could be in "casual dress."

---

[2]I was pleased to learn from reporters that in events where students were present or could watch a simultaneous podcast, my informal outfit was very much appreciated.

The festivities associated with awarding the Nobel Prizes extend over one week in December with the award ceremony itself always taking place on December 10, the anniversary of Alfred Nobel's death in 1896. It is the part of winter when the days are very short (sunrise is at 8:30 am; sunset at 2:50 pm) and having the Nobel Ceremony with all its festivities at that time of the year serve as a lively distraction for the people of Stockholm.

In his will, Nobel devoted a large part of the fortune he had made from the invention of dynamite to fund the Nobel Prizes, the first prizes being awarded in 1901. Although dynamite had many peaceful uses, it rapidly came to be employed in war. Nobel wrote in 1891 to his long-time confidant, Bertha von Sutter, about his hope "that on the day that two army corps can mutually annihilate each other in seconds, all civilized nations will surely recoil and disband their troops." [**League of National Archive, Geneva**]. Alas, this hope was not realized and the world had to wait for an invention ten million times more powerful than dynamite for a weapon that, at least so far, has been sufficiently destructive to help prevent world wars. It was, in part, as a result of his recognition of the destructive role of dynamite in war that he established the Peace Prize and the other prizes, all to be conferred "to individuals who contributed the greatest benefit for mankind." Bertha von Sutter, who was a peace activist, received the Nobel Peace Prize in 1905.

The Stockholm committee organizes the events in the Nobel week with great care. On November 7, we received a detailed program for our stay in Stockholm from our arrival on December 5 to our departure on December 14. The program included not only the formal activities (i.e., the lecture, the award ceremony, the Nobel banquet), but several others organized especially for me and my guests. Having been to Stockholm before, I had also arranged for private dinners with my family and friends at restaurants I knew, such as Wedholms Fisk, a classic fish restaurant (Figure 22.2).

An element that made our time in Stockholm almost magical was that we had a personal limousine and driver available at any time day or night from the moment of our arrival at the airport, where we were ushered into the VIP Lounge, throughout our stay in the luxurious Grand Hotel during the Nobel week, and finally for our VIP departure. Erik Anderson, our designated chauffeur, was special, both in personality and in the execution of his duties (Figure 22.3).

Figure 22.2. Family dinner at the Wedholms Fisk restaurant

Figure 22.3. Erik Anderson and Eva Nilson Mansfeld with Marci and me

Among other reasons, it turned out to be important for us that he was very familiar with the Stockholm restaurant scene. When we wanted to escape some of the formal receptions, we had only to ask him for a suggestion of a nice place to eat. One time he drove us from a reception given in the Nordic Museum to a

Figure 22.4. Street decorated for our passage

charming bistrot with excellent food. Apparently the restaurant was popular and fully booked, but Erik knew the owner and was able to reserve a table for Marci and me.

When we had been driven from the Grand Hotel to the Nordic Museum for the reception (Figure 22.4), the streets were lined with people who had flags and waved as we were going by. The police arranged for us to be able to drive without stopping, as if we were visiting dignitaries, which I suppose we were. In going to each event, Marci and I went with Erik in the limousine, while family and other invitees were transported in buses.

There were receptions held at several embassies. A notable one was an intimate and very personalized reception held at the Austrian ambassador's home. Among the guests was Professor Anton Zeilinger, who flew from Vienna especially to honor me for being awarded a Nobel Prize (see Chapter 21).

The one event that had required considerable preparation on my part was the Nobel Lecture on December 8 in the Aula Magna, a large lecture hall (Figures 22.5 and 22.6).

The major events were the Award Ceremony on the afternoon of December 10 in the Stockholm Concert Hall and the Nobel Banquet that evening in the City Hall.

Figure 22.5. Professor Sven Lidin introducing my Nobel Lecture

Figure 22.6. Family waiting to listen to Nobel Lecture

Both events required formal attire; that is, white tie and tails for me and a long evening gown for Marci, which she sewed herself (Figures 22.7 and 22.8).

The Award Ceremony itself is very elegant with all the Laureates in a row on stage in canonical order (Figure 22.9). Behind us on stage were previous Laureates who had come to Stockholm for the ceremony, as well as other dignitaries. The prize itself was awarded personally by King Carl XVI Gustaf of Sweden. Each prize

Figure 22.7. Marci and I in formal attire

Figure 22.8. Marci and I, with the children (Reba, Tammy, and Mischa) and grandchild (Rachel) before the Award Ceremony

Figure 22.9. Laureates on stage with King and Queen and the royal family

winner is introduced and comes forward toward the King, who rises to receive him/her. The appropriate medal and diploma are handed to the King, who in turn presents them to the Laureate (Figure 22.10).

Figure 22.10. Receiving the award from King Gustaf

As he shook my hand, I said thank you very much to him in Swedish (*tack så mycket*). This led him to start speaking rapidly to me in Swedish, of which I did not understand a word. Taking note of the situation, an official standing nearby whispered something to the King, who released my hand. Then, as we had been taught in the rehearsal in the Concert Hall, I walked backward away from the King, bowed to him when I reached the middle of the stage, then bowed to the dignitaries on the stage, followed by a bow to the audience, and my return to the line of Laureates. All of this sounds rather formal and contrived, but somehow it was an impressive event for me, my family, and everyone present.

The Nobel Banquet held in the City Hall is an extravagant affair with 1,300 guests (Figure 22.11). The entrance of the King and Queen, as they come down the steps into the hall is very formal (Figure 22.12).

The menu was beautifully designed, but the food itself was not quite up to our three-star restaurant standards. Realistically that would have been an impossible feat given the large number of people. However, the service was impeccable and rather astounding,[3] again given the number of guests.

---

[3] Young women and men from all over Sweden compete to serve at the Nobel Banquet. The ones selected are carefully trained, such that all 1,300 guests are served each course almost simultaneously.

Figure 22.11. The banquet

For someone who enjoys fine cuisine and lively discussion, such formal dinners can be stifling. The level of noise in this large setting meant that, at best, one could converse only with those sitting next to you; even talking with those sitting across the table was nearly impossible (Figure 22.13).

I had the good fortune to be seated next to Princess Christina, the King's sister, who was an interesting conversationalist, having among other things served as the chairwoman of the Swedish Red Cross (Figure 22.14).

Nobel Prize winners are strongly urged to visit a number of Swedish universities near Stockholm. One of my former postdocs, Kwangho Nam, had recently taken a position at Umea University, which is about 400 miles north of Stockholm. I decided to visit him even though Umea is outside the normal Nobel Prize "circuit." Moreover, I was tempted by the promise of a husky sled ride. Unfortunately, there was not enough snow so early in the winter for that. However, it was clearly something special for Umea University to have a Nobel Laureate come to give a lecture so that I was very pleased that I had gone. Since I spoke to a large audience, not only scientists, I presented my "Marsupial Lecture." Kwangho Nam, who like all my students, was aware of my interest in cooking, organized a dinner in a well-known, local farm house restaurant, Heds Gård. The chef had planned to

Figure 22.12. King Gustav and Queen Silvia entering the banquet hall followed by Princess Christina and me

work alone in preparing the dinner for our large party. He graciously accepted my offer to help in the kitchen, which made the dinner all the more special for me (Figure 22.15).

The Nobel Medal was designed by the Swedish sculptor and engraver, Erik Lindberg, and was first awarded in 1902 because it was not finished in time for the first Nobel Prize in 1901 (Figure 22.16).

Figure 22.13. Marci next to Peter Higgs

Figure 22.14. Princess Christina next to me at dinner

Figure 22.15. Me helping the chef at Heds Gård

Figure 22.16. Nobel Medal

It is solid gold and worth about $10,000. When we returned to Boston, I declared it to the customs agent who did not quite know what to do. Finally, he said I should ask the agent who checks you when you are ready to leave the arrival hall. I showed the medal to him and he proceeded to call over the other agents. Of course, I was worried that they were going to decide what duty to charge, as had happened when I imported more than the allowed number of bottles of wine. However, it turned out that he was so impressed by the medal that he wanted to show it to his buddies and then waved us through with a big smile.

Chapter 23

Science After the Nobel Prize Simulation

───────── ❧ ─────────

The Nobel Prize was awarded in 2013, 35 years after the BPTI molecular dynamics simulation on which it was based (see Chapter 15). Although both photography and cooking are activities that play an important role in my life, my primary focus continues to be on science and what I can do to contribute to the advancement of knowledge. In this chapter, I will outline how my research has evolved in a number of major areas since the first molecular dynamics simulation of a protein. Working to understand some of the systems involved has occupied me for much of my scientific life.

The chapter describes my research concerned with, among others, the mechanism of protein folding, the function of enzymes, the working of molecular machines, and the role of allostery in biology. An important development essential for certain of these studies involved improvements in free energy simulations methodology. In what follows, I shall indicate what has been learned from these studies, in some cases by citing one or more examples, in others by providing an overview. The research was done in collaboration with a group of more than 250 students, postdocs, and other coworkers, who have come to be called Karplusians (see Appendix 1 for a list). A symposium held in honor of my 75th birthday at the National Institutes of Health in April 2005 gives an indication of what some of the Karplusians have been doing in their research after they left my group [**Post and Dobson, 2005**].

Since the first molecular dynamics simulation of a protein molecule was published in 1977 [**McCammon, Gelin, and Karplus, 1977**], the field has had an explosive development. The original study of BPTI has stood the test of time and, more significantly, it served to open a new field [**Brooks, Karplus, and Pettitt, 1988**]. A large community of researchers is now active in the use of simulation methods to study the function of biomolecules. There are now a variety of computer programs, in addition to CHARMM [**Brooks et al., 1983; Brooks et al., 2009**], such as NAMD [**Phillips et al., 2005**], and GROMAX

[**Abraham** *et al.*, **2015**], that are generally available. Given that, unlike the early days of biomolecular simulations, the contributions of my group represent only a part of what is happening in this area of science.

As expressed in my Nobel Prize lecture (Figure 23.1), it continues to be my hope that molecular dynamics simulations will become a tool, like any other, to be used by experimentalists as part of their arsenal for solving problems. The simulated annealing method for determining X-ray structures originally proposed in 1987 by Axel Brünger, John Kuriyan, and myself [**Brünger, Kuriyan, and Karplus, 1987; Brünger and Karplus, 1991**], and so ably developed by Brünger in the program entitled XPLOR and the more recent version, CNS, is an application of simulations that is now an essential part of structural biology. However, a universal acceptance of molecular dynamics simulations methods for extending experimental data to learn how biomolecules function is still in the future [**Karplus and Lavery, 2014**].[1] Certainly, the recognition for simulations provided by the Nobel Prize has been helpful in encouraging their more widespread usage.

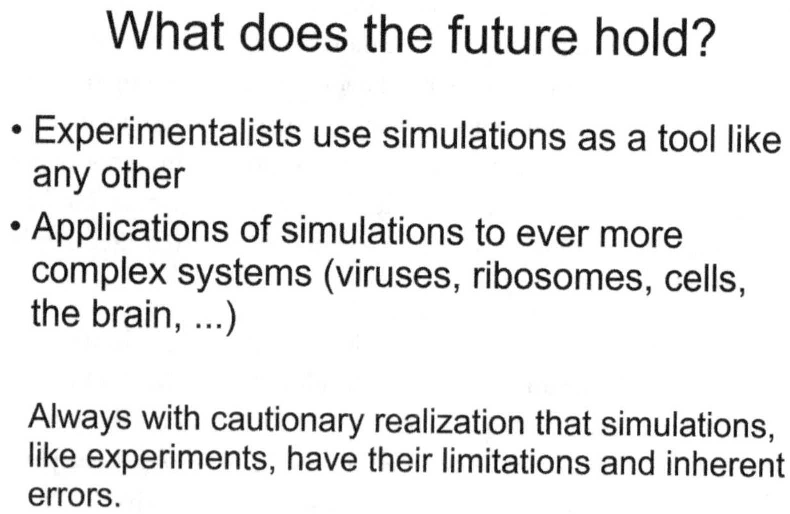

What does the future hold?

- Experimentalists use simulations as a tool like any other
- Applications of simulations to ever more complex systems (viruses, ribosomes, cells, the brain, ...)

Always with cautionary realization that simulations, like experiments, have their limitations and inherent errors.

Figure 23.1. Last slide of Nobel lecture

---

[1]In this paper, we point out that there were very few predictions made based on molecular dynamics simulations that were of sufficient interest to biochemists so that they undertook experiments to test them.

I still marvel at the insights that simulation methods can provide concerning the functions of biomolecules. Claude Poyart, a dear friend with whom I worked on hemoglobin [**Lee *et al.*, 1988**], characterized molecular dynamics simulations of proteins with a beautiful image. He likened the X-ray structures of proteins to a tree in winter, beautiful in its stark outline but lifeless in appearance. Molecular dynamics gives life to this structure by clothing the branches with leaves that flutter in the thermal winds.

Most of the motional phenomena examined during the first ten years after the BPTI simulation paper was published continued to be studied both experimentally and theoretically. The increasing scope of molecular dynamics due to improvements in methodology and the tremendous increase of the available computer power has made possible the study of systems of greater complexity on ever-increasing timescales. Some of the results from such studies are described in an issue of *Accounts of Chemical Research* [**35, 2002**], entitled "Molecuar Dynamics Simulations of Biomolecules" for which I was the guest editor. I am particularly proud of my introductory editorial, because I was able to persuade the editors to include one of my favorite paintings *The Birth of Venus* by Sandro Botticelli (Figure 23.2). The point of showing the picture was that unlike Venus, who rose from the sea fully formed in all her glory, such creations rarely occur in science, particularly today when every subject is being studied by many people.

Figure 23.2. The Birth of Venus

Two attributes of molecular dynamics simulations have played an essential part in the explosive growth in the number of studies based on simulations. As already mentioned, simulations provide the ultimate detail concerning individual particle motions as a function of time. For many aspects of biomolecule function, these details are of importance. For example, in myoglobin, which stores oxygen in muscles, they show the pathways by which oxygen enters into and exits from the heme pocket. The other important aspect of simulations is that, although the potentials employed in simulations are approximate, they are completely under the user's control. By removing or altering specific terms in the potential function, their role in determining a given property can be examined. This is most graphically demonstrated in calculations of free energy differences, which are based on "computer alchemy," in which the potential is transmuted from that representing one system to another during the simulation. I introduced the term "computer alchemy" for this method because it is analogous to what the ancient alchemists claimed to do in transmuting lead into gold, something easy to do on a computer [**Gao *et al.*, 1989; Simonson, Archontis, and Karplus, 2002; Wong and McCammon, 1986**].

There are three types of applications of simulation methods in the macromolecular area, as well as in other areas involving mesoscopic systems. The first uses the simulation simply as a means of sampling configuration space. This is involved in the utilization of molecular dynamics, often with simulated annealing protocols, to determine or refine structures with data obtained from experiments. The second uses simulations to determine equilibrium averages, including structural and motional properties (e.g., atomic mean-square fluctuation amplitudes) and the thermodynamics of the system. For such applications, it is necessary that the simulations adequately sample configuration space, as in the first application, with the additional condition that each point be weighted by the appropriate Boltzmann factor. The third application employs simulations to examine the actual dynamics. Here not only is adequate sampling of configuration space with appropriate Boltzmann weighting required, but it must be done so as to properly represent the time development of the system. Monte Carlo simulations, as well as molecular dynamics, can be utilized for the first two applications. By contrast, in the third application, where the motions and their time developments are of interest, only molecular dynamics can provide the necessary information.

Studies of Protein Folding

My interest in protein folding has continued since my sabbatical with Shneiror Lifson's group in 1969 where I saw Chris Anfinsen's cartoon that was the inspiration for the diffusion–collision model developed with David Weaver (see Chapter 14). Since then there have been more than thirty papers concerned with protein folding from my group, of which the most recent appeared in 2014 [**Ovchinnikov and Karplus, 2014**].

When David and I developed the diffusion–collision model in 1976 [**Karplus and Weaver, 1976**], protein folding was a rather esoteric subject of interest to a very small community of scientists. The field has been completely transformed because of its importance for understanding the large number of protein sequences available from genome projects and because of the realization that misfolding can lead to a wide range of human diseases [**Dobson, 2003**]; these diseases are found primarily in the older populations that form an ever-increasing portion of humanity.

Scientists, both experimentalists and theoreticians—physicists, as well as chemists and biologists—now study protein folding. Over the past decade or so the mechanism of protein folding has been resolved, in principle. It is now understood that there are multiple pathways to the native state and that the bias on the free-energy surface, due to the greater stability of native-like versus nonnative contacts, is such that only a very small fraction of the total number of conformations is sampled in each folding trajectory [**Dobson, Sali, and Karplus, 1998**]. This understanding was achieved by the work of many scientists, but a crucial element was the study of lattice models of protein folding. Such toy models, as I like to call them, are simple enough to permit a sufficient number of trajectories to be calculated to make possible a meaningful statistical analysis of the folding process and free-energy surface sampled by the trajectories [**Sali, Shakhnovich, and Karplus, 1994**]. However, they are complex enough so that they embody the Levinthal "paradox" [**Karplus, 1997**], i.e., there are many more configurations than could be visited during the calculated folding trajectory (Figure 23.3).

Although each folding trajectory is different, the bias to the native state on the free energy surface is sufficient that the stochastic search samples only a small fraction of the total configurations in finding the native state. This principle is

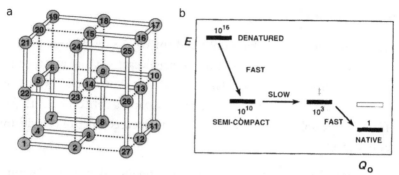

Figure 23.3. (a) An example of a compact self-avoiding structure of a chain of 27 monomers (filled numbered circles) with 28 contacts (dashed lines). (b) Three-stage mechanism of folding of the present model; the empty rectangle indicates the energy of the native state of a non-folding sequence

now generally accepted as the solution of the search problem in protein folding [Karplus, 1997; Wolynes, 2005]. As computer power increased, it became possible to shift from the lattice models to simulations of realistic descriptions of proteins. The objective of present studies are concerned with the specifics of the folding of individual proteins and the ways that misfolding is avoided [Lindorff-Larsen et al., 2011].

It is important to note that the implication of these early studies was in part psychological. Even though the lattice models use a simplified representation, actual folding was demonstrated on a computer for the first time. An article by me based on a lecture at a meeting in Copenhagen [Karplus, 1997] describes this change in attitude as a paradigm of scientific progress.

> Changes in perception are an essential element in the advancement of science. Often, the key to progress is not that a given view has been disproved and that the view that replaces it has been proved. Instead, the important element is the acceptance of the new view by the scientific commmunity. That is what appears to be happening in the case of the protein folding problem. It is not a great exaggeration to say that the pessimistic viewpoint of yesterday ('It is impossible for proteins to fold to the native state, even though they do so readily in solution') has been replaced by the optimistic viewpoint (the "new view") of today ('It is obvious that proteins should be able to fold rapidly to the native state').[2]

---

[2] The widespread attention to our *Nature* 1994 paper was due, in part, to a serendipidous event. Many readers of *Nature* read only the "News & Views" articles to find out what is new. The omission in the "News & Views" article by Baldwin [Baldwin, 1990] of references to the work of the many other people (e.g., Bryngelson and Wolynes, 1989; Camacho and Thirumalai, 1993; Dill et al., 1995; Go, 1983; Harrison and Durbin, 1985) who contributed to developing the "new view" led to a personal, as well as a published, correspondence [Chan, 1995; Karplus, Sali, and Shakhnovich, 1995] that appears to have contributed to the widespread dissemination of the conclusions.

Given the increase in the speed of computers, particularly through the development of massively parallel machines, it is likely that within the next ten years *ab initio* folding will reach the stage where it can be used not only to determine the details of the folding mechanism, but also to aid in solving the "other" folding problem, that of determining the structure of the native state directly from the sequence by molecular dynamics simulations. In fact, a special purpose supercomputer, named ANTON,[3] developed by David Shaw and his group [**Shaw et al., 2010**] has been successful in repeatedly folding small proteins [**Lindorff-Larsen et al., 2011**].

On a personal note, I must add that this type of brute force approach, in which the computer churns out the answer, is of less interest to me than solving the conceptual problem of protein folding, which has been accomplished by more approximate techniques, as described above [**Dobson, Sali, and Karplus, 1998**].

Enzyme Catalysis

Enzymes are nature's catalysts. They speed up many of the chemical transformations required by living systems. Some enzymes can produce rate accelerations of a factor of 10^{19} [**Wolfenden and Snider, 2001**]. The rate constant k of a reaction can be written in the Arrhenius form $k = A(T) \exp(-\triangle G^{\ddagger}/RT)$, where $\triangle G^{\ddagger}$ is the activation free energy and A(T) the pre-exponential factor. In 1946, before structural information about proteins was available, Linus Pauling proposed [**Pauling, 1946**] that enzymes can accelerate reaction rates because they bind the transition state better than the substrate and thereby lower $\triangle G^{\ddagger}$. The validity of this key concept in enzyme catalysis has been confirmed by many studies [**Garcia-Viloca et al., 2004; Schowen, 1978**], although dynamical contributions to A(T) are of importance in some cases [**Garcia-Viloca et al., 2004**].

Studies to understand how enzymes catalyze reactions have occupied me for many years and the methods used are a prime example of applications in the rubric of the Nobel Prize citation ("for the development of multiscale models for complex chemical systems"). In fact, an essential element of progress in this area has been the realization that it is necessary to simplify the calculations by treating

---

[3]The name ANTON was chosen to commemorate Anton van Leewenhoek, the inventor of the microscope, who used it to observe muscle fibers, bacteria, spermatazoa, and blood flow in capillaries.

the "core" potential energy surface involved directly in the reaction by one or another quantum mechanical approach and by describing the environment (the rest of the protein plus solvent) by the empirical potential surface used in molecular dynamics simulation of nonreactive systems. We note here that in most cases (not all) the dynamics of the reaction can be adequately described by classical mechanics, though some reactions, such as H-atom transfers, require quantum mechanics [**Garcia-Viloca et al., 2004**].

Computer simulations have played an important role in relating the observed lowering of the activation free energy to the enzyme structure and its flexibility. The environment provided by specific amino acids is such that it produces the greater stabilization of the transition state, relative to the reactant state, as proposed by Pauling. A certain degree of enzyme flexibility is also essential for catalysis since atomic motions of the enzyme are required for the reactions to take place [**Brooks, Karplus, and Pettit, 1988**]. Larger scale conformational changes are often involved as well. In many cases, the catalysis begins with an open structure that allows the substrate to enter prior to the reaction, followed by a closing motion to form a reaction "chamber" isolated from solvent, and finally an opening after the reaction that allows the products to escape [**Henzler-Wildman et al., 2007**] (Figure 23.4).

Although the original idea for combining quantum mechanical and classical mechanical calculations goes back to 1971 [**Honig and Karplus, 1971**], it had to wait nearly twenty years for a general formulation [**Field, Bash, and Karplus, 1990**], which came to be called the QM/MM methodology. This made possible calculations with sufficient accuracy to obtain meaningful results for enzyme mechanisms. In what follows, I will describe a set of my studies that I think played an important role in the field. The reader should remember that in "my studies" essential collaborators, mainly students and postdocs, were involved.

I first applied the QM/MM methodology to triosephosphate isomerase (TIM), the enzyme that catalyzes the transformation of dihydroxyacetone phosphate (DHAP) to (R)glyceraldehyde 3-phosphate (GAP) (Figure 23.5).

This reaction is part of the glycolytic pathway, which generates ATP, the energy currency of the cell (see section "Molecular Motors").

Adenylate Kinase

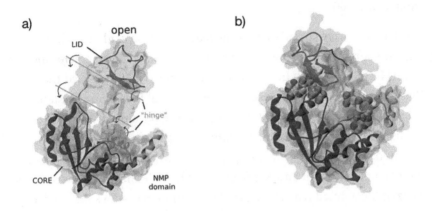

$$2A\text{-}P\text{-}P \;\rightleftharpoons\; A\text{-}P\text{-}P\text{-}P + A\text{-}P$$

Figure 23.4. Adenylate Kinase. (a) Open structure in the absence of substrate. (b) Closed structure with two molecules of the substrate, adenosine diphosphate (A-P-P), before the reaction

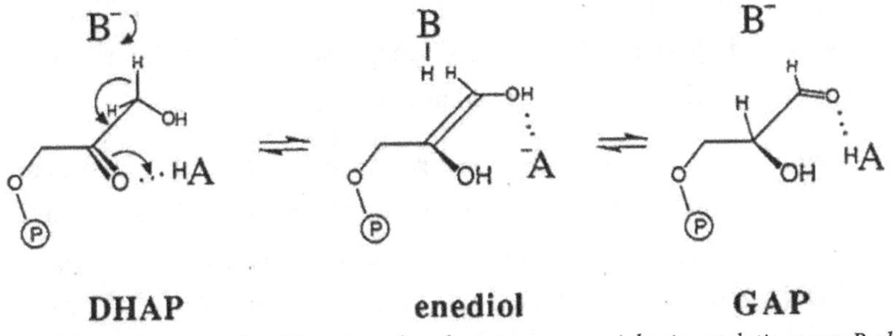

DHAP   **enediol**   **GAP**

Figure 23.5. The reaction catalyzed by triosephosphate isomerase. A basic catalytic group B abstracts the pro-R proton on C-1 of dihydroxyacetone phosphate (DHAP), assisted by as acidic catalytic group HA, to produce the intermedate enediol. This intermediate then collapses to give the product glyceraldehyde 3-phosphate (GAP) and regenerates the enzyme [From Knowles, J. R. (1991) Nature 350 pp. 121–124]

TIM has been studied in detail by many experiments and simulations. The apparent barrier for the reaction in the enzyme has been shown to be 11–13 kcal/mol lower than that for the reaction in aqueous solution [**Cui and Karplus, 2002; Feierberg and Aqvist, 2002**], leading to a rate of acceleration of about 10^{10}. It is the first enzyme to be called "perfect" in the sense that the reaction itself is accelerated so much that it is no longer the rate-limiting step, which in the case of TIM is the binding of the substrate [**Albery and Knowles, 1976**]. In other "perfect" enzymes, such as adenylate kinase, it is the release of products [**Henzler-Wildman et al., 2007**].

Two mechanisms were considered for the reaction [**Bash et al., 1991**]. In one, His 95(+) is the source of the proton to form the enediolate intermediate, and in the other, the His 95 is neutral. In the Bash *et al.* paper it was proposed that neutral His 95 is the donor because the use of His 95(+) would result in overstabilization of the intermediate, which would trap the reaction in a deep well, preventing it from going to completion. This surprising prediction (the dogma was that only His(+) could serve as the source of a proton) was confirmed in a subsequent experimental paper [**Lodi and Knowles, 1991**]. Since then it has been found in other enzyme-catalyzed reactions. TIM has continued to be a subject of research in my group and the latest paper was published in 2003 [**Cui and Karplus, 2003**].

TIM is also of interest because it is a striking example of the conformational change that can occur to protect the active site from side reactions by water. By evolution, TIM developed a structure with a relatively rigid 11-residue loop with "hinges," which allows it to fold to protect the active site like a "lid," once the substrate is bound [**Joseph, Petsko, and Karplus, 1990**]. In this sense, TIM is similar to adenylate kinase, as discussed above.

Since the original simulation of the TIM reaction in 1991, my coworkers and I have studied many other enzymes, often in collaboration with experimentalists. A set of such collaborative studies, in this case with the group of Greg Verdine, concerned DNA repair enzymes. Because of the importance of DNA as a repository of genetic information, there are many enzymes that repair errors due to damaged nucleobases. What is involved is the surprising ability of detecting one modified base in a million-fold excess of undamaged DNA. The problem in this case arises from the oxidation of guanine (G), one of the four DNA bases, to oxoguanine (oxoG) [see (Figure 23.6) **Qi et al., 2009**].[4]

Greg Verdine and his group were studying how the bacterial enzyme MutM detects oxoG and extrudes it from the DNA double helix for excision. When

---

[4]5-Fluorouracil is an interesting case of a modified base being used to treat skin cancers and warts. Incorporation of the modified base prevents reproduction of the cell. Since cancer cells or wart cells reproduce much more rapidly than normal cells, they are affected first. A limitation of 5-fluorouracil is that it can only be used externally. If it were ingested, it would also prevent normal cell reproduction.

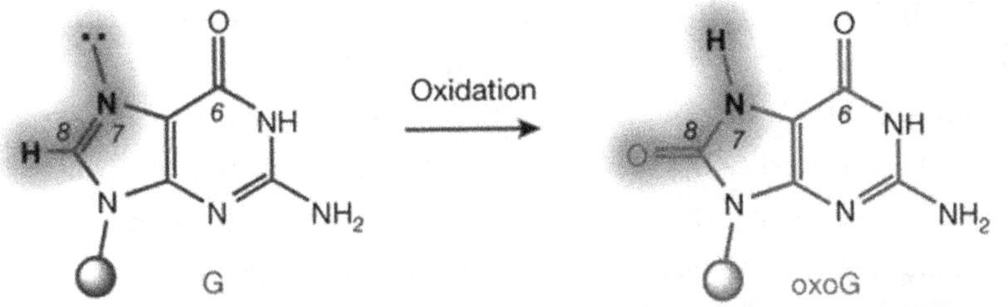

Figure 23.6. Structural comparison of G versus oxoG with the difference highlighted

Verdine submitted the paper to *Nature* the editors responded by indicating the paper was of interest, but would be accepted only if there was a computer simulation to show the pathway by which the extrusion takes place. This unexpected condition for publication, the first to my knowledge, was a pleasant surprise. It led to a fruitful continued collaboration between the two groups [e.g., **Nam, Verdine, and Karplus, 2009**].

I had earlier studied another DNA repair enzyme, uracil-DNA glycosylase with Aaron Dinner, in a collaboration with G. M. Blackburn. We concluded that it had the surprising attribute that it acts by substrate autocatalysis; that is, the substrate itself, as well as the enzyme, contributes to lowering the activation barrier [**Dinner** *et al.*, **2001**]. This prediction led to a series of interchanges with James Stivers, who did experiments specifically designed to test our prediction and concluded that there was a disagreement [**Jiang** *et al.*, **2003**]. In a subsequent paper, which was based on extensive simulations [**Ma** *et al.*, **2006**], we showed how the apparent discrepancy could be resolved.

Molecular Motors

Molecular motors are some of the most remarkable proteins found in living systems. In fact, cells have been described as a "collection of machines" [**Alberts, 1998**]. The motors generally are enzymes that increase the rate of the reaction involved in their function. However, unlike ordinary enzymes, they make use of the reaction to do work. Nature has "designed" these machines by evolution (i.e., by specifying the amino acid sequence of the proteins that make up the motor) to function by ligand-induced conformational changes in their structure. Because of their great intrinsic interest and the need for simulations, as well as experiments, to understand how they work, several of them have been the subject of my research over the years. One group of such motors are the myosins, which in addition to their role in the contraction of muscles [**Kuriyan, Konforti, and Wemmer, 2017**], also have family members (e.g., myosin V) that transport material from one part of the cell to another along actin fibers. Another group of motors are the kinesins, which also transport material, but instead of actin they use microtubules as the "tracks" on which they move. The kinesin that Wonmuk Hwang, Matt Lane, and I studied by molecular dynamics simulations [**Hwang, Lang, and Karplus, 2017**] consists of two globular "feet" and a stem, which forms a coiled-coil, at the top of which is the container for the material that is transported. It is now known that some kinesins "walk" on the microtubules, while others jump, and some destroy the microtubule as they move along it. One kinesin plays an important role in cell division. Since cancer cells divide faster than normal cells, inhibitors of this kinesin are being studied as possible anticancer drugs [**Huszar *et al.*, 2009**]. Some viruses, e.g., the smallpox virus, are able to attach themselves to the kinesin container, and can thereby be transported from one end of the cell to the other on the microtubule track in minutes, instead of the ten or so hours that would be required for diffusion in the concentrated cellular medium.

The energy source for the movement of kinesins and many other motor proteins is ATP (adenosine triophosphate). It has been called the energy currency of the cell because it stores bond energy that is available when needed for a wide range of cellular functions. The motor protein, F_oF_1 ATP synthase, regenerates ATP from its hydrolysis products ADP and Pi ($H_2PO_4^-$). It is found in eubacteria, chloroplasts, and mitochondria and is responsible for most of the ATP synthesis in living systems (Figure 23.7).

Unlike linear motors like myosin and kinesin, ATP synthase is a rotatory motor as demonstrated by single molecule experiments [**Noji *et al.*, 1997**], published in the same year that Boyer and Walker received the Nobel Prize for their contribution to the understanding of F_oF_1 ATP synthase. When I first saw the Noji *et al.* movie of the rotating γ-subunit, it was in a certain sense an epiphany for me. I felt that if molecular dynamics simulations were good for anything, they should be able to explain how this "wonderful machine," in the words of Boyer, works.

The F_o component, which is mainly in the membrane, converts a proton gradient across the membrane into rotational motion of a drive shaft (the γ-subunit of F_1;

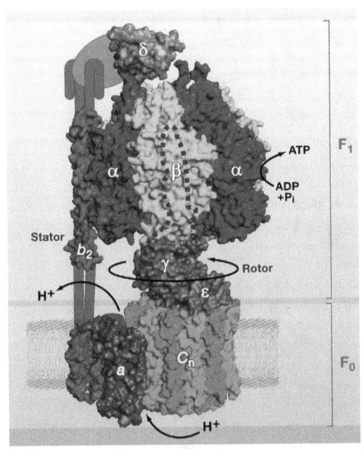

Figure 23.7. The ATP synthase of Escherichia Coli. *The rotor consists of* γ, ε, *and a ring of c- subunits (10 in E. coli; 10–15 in other organisms). The dotted line indicates the helical coiled-coil extension of the* γ-*subunit into the central cavity of the* $\alpha_3\beta_3$ *hexagon. Three catalytic sites are located at* α/β *interfaces. The "stator" (ab₂δ) prevents co-rotation of catalytic sites with the rotor. Protons traveling down the transmembrane* H^+ *gradient, between a and c, generate rotation of the c ring, making the* γ-*subunit also rotate, which drives ATP synthesis by forcing different conformations sequentially of each of the catalytic sites. For ATP hydrolysis-driven proton extrusion, all arrows are reversed. ATP synthases from chloroplasts and mitochondria follow the same functional principles, but show greater subunit complexity [Reproduced with permission from Senior, A. E. (2007) Cell 130 pp. 220–221]*

see Figure 23.7). The rotation of the γ-subunit induces conformational changes in F_1, which result in the synthesis of ATP. Boyer's proposal, made in advance of structural information [**Boyer, 1993**] that F_0F_1 ATP synthase is a rotary motor was strongly supported by the first crystal structure of F_1 ATPase [**Abrahams et al., 1994**]. The intracellular globular portion works in reverse in the presence of ATP. It hydrolyzes ATP to ADP and Pi to rotate the γ-subunit.

Our first paper [**Ma et al., 2002**], gave some insight into the function of the motor. It was published in 2002, in collaboration with Andrew Leslie and John Walker, whose X-ray structure of F_1 ATPase was essential as a basis for our simulations. That paper represented the limits of molecular dynamics methodology at the time and, as is sketched in what follows, experiments and simulations of F_1 ATPase have continued to this day.

The central rotating shaft (the γ-subunit) is surrounded by three noncatalytic α-subunits and three catalytic β-subunits. Importantly, three conformational states adopted by the β-subunits were found in the original crystal structure [**Abrahams et al., 1994**], and molecular dynamics simulations gave the first insights into how the conformations change as a function of the γ-subunit orientation [**Böckmann and Grubmüller, 2002; Ma et al., 2002**]. It was not possible from the observed conformations, *per se*, to determine which were the high-, medium-, and low-affinity states for MgATP binding that correspond to the three binding constants measured in solution [**Weber et al., 1993**]. Associating the measured binding constants with the conformational states was an essential step for developing a molecular model for the motor function [**Gao, Yang, and Karplus, 2005**]. In 2003, free energy calculations, of the type that are now referred to as "computer alchemy" (see section "Free Energy Simulations"), suggested that the binding site in the so-called β_{TP} conformation was the high affinity site [**Yang et al., 2003**]. This was confirmed four years later by a FRET experiment done specifically to test simulation results [**Mao and Weber, 2007**].

One reason that simulations have made contributions to our understanding of the mechanism of F_1 ATPase is that the structures from X-ray crystallography are "silent" about the transition paths from one state to the other, and the forces involved. However, since the time for one rotation of the γ shaft is in the millisecond range [**Kinosita, Adachi, and Itoh, 2004**], it takes much longer than the submicrosecond timescale accessible to standard molecular dynamics simulations for such large systems; F_1 ATPase in its box of water consists of about 200,000 atoms, of which 150,000 are the water molecules. Because the beginning and end states corresponding to a 120° rotation of the γ-subunit are known [**Abrahams et al., 1994**], the challenge is essentially one of finding a low energy path that connects one stable conformation of the system to another, a general problem that continues to receive considerable attention [**Bolhuis et al., 2002**]. Details concerning the conformational transitions in F_1 ATPase were obtained by molecular dynamics simulations in the presence of biasing forces. The spirit in which these simulations are done is to "push" on the molecular assembly (e.g., by forcing the γ-subunit to rotate) and to analyze how the rest of the structure responds. Since the time scale of the forced rotational transition of the γ-subunit is orders of magnitude faster than the actual rotation rate, the implicit assumption in such studies is that meaningful information concerning the mechanism can still be obtained. These studies demonstrated that the rotation of the γ shaft triggers the opening of the nucleotide-bound β_{TP} subunit and the closure of the open β-subunit.

With these results and experimental solution kinetic constants identified with the different β-subunits, a detailed kinetic model for ATP hydrolysis by F_1 ATPase was developed [**Gao et al., 2003; Gao, Yang, and Karplus, 2005**]. However, certain revisions of this model are required, as shown by the more recent simulations of F_1 ATPase [**Nam, Pu, and Karplus, 2014**][**Nam and Karplus, 2019**]. Thus, a complete understanding of this "wonderful machine" is coming into view.

Role of Allostery in Biology

I became interested in allostery in the early 1970s when I heard a lecture by Max Perutz (see Chapter 12). He showed that human hemoglobin existed in two quaternary structures (deoxyT, T_0 and oxyR, R_4), in agreement with the MWC model of Monod, Wyman, and Changeux [**Monod, Wyman, and Changeux, 1965**]. As described there, Attila Szabo and I developed a statistical mechanical description based on the Perutz results. In 2008, Qiang Cui and I wrote a review [**Cui and Karplus, 2008**] entitled "Allostery and cooperativity revisted" stimulated to do so, in part, by the resurgent interest in allostery and, in part, by the publication of papers, mistakenly from my perspective, that implied that a "new view" of allosteric transitions had been discovered. This was referred to as the "population-shift" model (see, e.g., **Kern and Zuiderweg, 2003**). The authors neglected the fact that a "population shift" is the basis of the original MWC model, and therefore can hardly be considered new. To make this clear we quote [**Changeux and Edestein, 2005**]: "A critical statement of the MWC theory was that, in essence, the conformational transition that links the multiple sites present on the allosteric oligomer and mediates signal transduction involves states that are populated in the absence of ligand and may spontaneously interconvert with each other." Arguably, the most important aspect of the "new view" is that it has revived an interest in allostery, which had been a neglected field in biophysics for some years. (As Francis Crick is supposed to have remarked long ago, "Hemoglobin has a 'bore' effect.")

One aspect of the recent discussions concerns a question that arises in the population-shift model, namely, does the ligand bind to the conformation that has only a small population in the unliganded state and thereby shifts the equilibrium to that conformer, or does it shift the equilibrium by binding to the conformer that is dominant in the unliganded state? Hemoglobin is one of the few systems for which data were already available to test this conjecture. The equilibrium between T_0 and R_0 and the kinetic rate constants for ligand binding to the T_0 and R_0 states had been measured [**Eaton et al., 1999**]. They show that binding of the first ligand is to the low-affinity state (T_0) and not to the high-affinity state (R_0); the fourth oxygen does bind to the R state because that is now the predominant species.

The hemoglobin tetramer, composed two α-subunits and two similar β-subunits, can be considered to be composed of two $\alpha\beta$ dimers. The quaternary transition from the T_0 state to the R_4 state, based on the shortest path between the X-ray structures of the states, was described by Baldwin and Chothia [**Baldwin and Chothia, 1979**] as involving a 15° rotation of one "dimer" ($\alpha_1\beta_1$) relative to the other "dimer" ($\alpha_2\beta_2$) around a virtual axis, together with a small relative translation of the two dimers toward each other. This description of the transition was accepted until 2011, when Stefan Fischer, Kenneth Olsen, Kwanho Nam, and I [**Fischer et al., 2011**] made a study of quaternary transition based on the realization that a knowledge of two end structures does not provide direct information on the pathway connecting them.

Instead of the simple dimer rotation model of Baldwin and Chothia, the path was found to be more complex and to consist of two sequential quaternary rotations involving different rotation axes (Figure 23.8).

As noted in the paper, the research was essentially completed in 2000, but I was told by the editors of several journals that the manuscript would not be considered unless there was experimental evidence to support the proposed "iconoclastic" results. After some crucial experiments [**Cammarata et al., 2010; Spiro and**

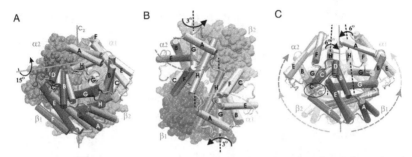

Figure 23.8. *The T→R transition of hemoglobin. (A) Comparing the T (i.e., deoxy, shown in color) and R (tetraoxy, in gray) conformers (superposed by fitting their $\alpha_2\beta_2$-subunits). The $\alpha_1\beta_1$-dimer as a whole appears rotated by approximately 15° (the screw axis is shown in cyan). The C2-symmetry axis relating the $\alpha_1\beta_1$- and $\alpha_2\beta_2$-dimers is in magenta. (B) First quaternary event (Q1): relative to its position in the T state (in color), each α-subunit rotates by 3° around its G helix (the two α-subunits pivoting toward each other), resulting in the intermediate halfway along the path (in gray). The view is down the central channel, along the C2-symmetry axis ($\otimes$). (C) Second quaternary event (Q2): relative to its position halfway along the path (in color), each $\alpha\beta$-dimer rotates by 6° (R state shown in gray) around its αH helix. One of the two switch regions (α_2C/β_1FG loop) is circled*

Balakrishnan, 2010] validating the results were published in 2010, we made a direct submission to the PNAS where the article was accepted and published. This succession of events indicates that the difficulty of publishing predictions from simulations in biology, mentioned earlier (see Chapter 12), had not disappeared. Hopefully the Nobel Prize is helping to change the situation.

In 2012, I gave a lecture at a symposium in Rome, organized to honor Maurizio Brunori on his 75th birthday. It was entitled "Hemoglobin Forever: Yesterday, Today, and Tomorrow." At the time, I reported on the results of our research mentioned above, as well as some others [**Zheng et al., 2013**]. I did not think that the "Tomorrow" part of my title meant that I would still be working on hemoglobin as I am writing this today.

Free Energy Simulations

Since the free energy of biological macromolecules as a function of their co-ordinates is a quantity essential for understanding their properties, my research in this area goes back to some of the earliest publications. An example is the 1981 paper with Joseph Kushick on the configurational entropy of macromolcules

[Karplus and Kushick, 1981]. Since then there have been numerous papers on the subject of free energy simulations, but rather than presenting many of them, I will describe two reviews. The first is in the 2002 issue of *Accounts of Chemical Research* **[Karplus, 2002]** entitled "Molecular Dynamics Simulations of Biomolecules," of which I was a guest editor. In addition to the Preface, I wrote one of the articles with coauthors Thomas Simonson and Georgios Archontis entitled, "Free Energy Simulations Come of Age: Protein Ligand Recognition" **[Simonson, Archontis, and Karplus, 2002]**. As pointed out there, the developments in the free energy simulation methodology and the increase in computer power are such that the calculation of the free energy of binding of a small molecule to a protein gives meaningful results, i.e., they are sufficiently accurate to be used in the virtual screening of drug candidates. The second was prepared thirteen years later when the journal *Molecular Simulations* published an entire issue devoted to free energy simulations. In the Preface to that issue **[Karplus, 2016]**, I wrote:

> Over the intervening thirteen years, developments have taken place that have made possible significantly improved free energy simulations. Moreover, they have become more widely used, if for no other reason than for the imprimatur from the 2013 Nobel Prize in Chemistry **[Karplus, 2014b]**.

Computers of significantly increased power have played an essential role in the improvements. Interestingly, 2015 was the 50th anniversary of Moore's Law **[Moore, 1965]**, which stated that computers double in speed every years based on data for only four years. When asked about the law that year, Moore said that he thought the doubling in speed might continue for ten years. His focus was on the transistor industry, and it amazed him that the law seemed to be still valid. In 2010, Michele Vendruscolo and Christopher Dobson **[Vendruscolo and Dobson, 2010]** described a law similar to Moore's for biomolecular simulations, which stated that their speed appeared to double every year. Given that, the computer power accessible to free energy simulations today is nearly 10,000 times greater than that in 2002. Now that microsecond, and even millisecond, simulations are becoming relatively routine, clear improvements in the statistical precision of free energy simulations are possible. Most of the papers in the special issue take advantage of this aspect, but it is not their focus. Instead, the papers are primarily concerned with developments in free energy methodology and its applications.

What Does the Future Hold?

The last slide of my Nobel lecture was entitled "What Does the Future Hold?" (Figure 23.9). I expressed the hope that experimental structural biologists, who know their systems better than anyone else, would make increasing use of molecular dynamics simulations for obtaining a deeper understanding of particular biological systems. When molecular dynamics is a routine part of structural biology, it will become clearer what refinements and extensions of the methodology are most needed to perfect the constructive interplay between the simulations and experiment. These challenges for simulation experts will hopefully catalyze new developments in the field.

I also stated in the slide that one can expect "Applications of simulations to cover more complex systems (viruses, ribosomes, cells, the brain, ...)." Some of these are already in progress, mostly by other researchers, many of whom were my students and postdocs at one time and now are professors in their own right. However, there is one area, the brain of a little worm, *Caenorhabditis elegans* (Figure 23.10) in which I have recently made a contribution [**Klein *et al.*, 2017**]. This worm has a very special nervous system, which has made it of particular interest to neuroscientists. Our brains are affected by our genes, but the development is not completely determined by them. This is made clear by the fact that two identical twins, who have the same genetic makeup, can behave very differently.

What does the future hold?

- Experimentalists use simulations as a tool like any other
- Applications of simulations to ever more complex systems (viruses, ribosomes, cells, the brain, ...)

Always with cautionary realization that simulations, like experiments, have their limitations and inherent errors.

Figure 23.9. What does the future hold?

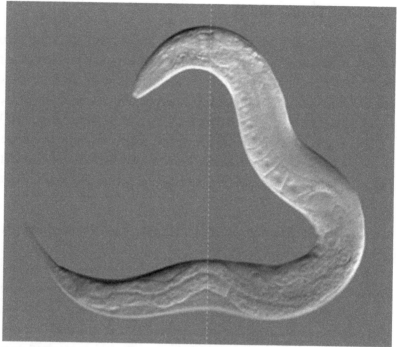

Figure 23.10. Image of Caenorhabditis elegans *showing its graceful form*

By contrast, *C. elegans* has a nervous system that is completely determined by genetics. Thus, except for rare mutations, the "brains" of all worms are identical. This makes possible measurements of the motions of a large number of them and to combine the data from the measurements to obtain statistically significant information.

What made the study of *C. elegans* of particular interest to me was the realization that analyzing their motions could be based on techniques that my coworkers and I had developed for studying the complexities of protein folding. Here we showed [**Klein *et al.*, 2017**] that the navigational dynamics of *C. elegans* has parallels with the complex dynamics of a polypeptide chain "navigating" to find its native protein structure. Both involve biased random search processes: the protein needs to find its native structure and *C. elegans* needs to find favorable conditions, such as a source of food or the right temperature. Neither search can be purely random, because it would not be effective. For the protein folding case, it would lead to the Levinthal paradox, which is avoided by a bias toward the native structure provided by the potential energy. Analogous considerations apply to the navigation dynamics of *C. elegans*. A purely random search would be inefficient due to the large size of the space accessible in their normal environment. Consequently, the worms use cues to bias their search. An example is a temperature gradient which plays the "role" of the potential energy. A purely deterministic search would also not be effective for proteins or worm, because there can be traps (local minima) in the accessible space. These minima could have a physical origin or be due to a complex non-monotonic nature of the cues. A stochastic component in the biased search allows *C. elegans* to overcome the trapping problem. The actual details of their navigational dynamics are specified by the neural circuitry that enervates the muscles. This has evolved through evolution, in analogy to the amino acid sequence in proteins, which is determined by evolution as well.

An exciting area in which I am working as I write this in 2019 has to do with vaccine development. The principle of vaccination is that it simulates an individual to develop antibodies against one or more virus strains. One gets an influenza shot every year because the flu virus is highly variable. The yearly vaccine is developed based on the best estimate in the spring of what the virus will be like in the coming fall. Sometimes, the estimate is not very good, as was the case for 2017–2018 flu season, when the vaccine turned out not to be very effective against the particularly virulent strain that reached epidemic proportions. It would be of great benefit to humankind if there were an influenza vaccine that would provide nearly permanent immunization against all forms of the virus that develop over the years. There are a few people who naturally have such permanent immunity. This suggests that it should be possible to find a vaccine to induce people to develop antibodies required for permanent immunity. Such antibodies are called "broad-based" antibodies because they work against many viral strains.

What my group was working on, in collaboration with Arup Chakraborty at MIT, is a related problem, that of developing a vaccine to prevent HIV from developing into AIDS. It is known that a small number of infected individuals, who have been kept alive without developing AIDS by the various antiviral drugs that are now available, became immune to the virus. They have developed the broad-based antibodies mentioned above by trial and error over time, often after ten years or so. The function of a HIV vaccine would be to induce the immune system to develop such broad-based antibodies, rather than having to wait for them to arise naturally. If our program of calculations, supported by experimental tests of the group of Dennis Burton, one of the world's experts on HIV, were to lead to a successful HIV vaccine, that would be, in a certain sense, a culmination of the methodology I began to develop forty years ago. If there is a chance of success, it is something I feel compelled to try.

Although I would be very disappointed if we did not succeed in developing a vaccine, it would not affect my career. However, for the students and postdocs involved, success is very important for their future. Given that, I am making sure that they at least have publications that describe the scientific developments required for the vaccine project. One of them is the paper by Victor Ovchinnikov and coworkers [**Ovchinnikov et al., 2018**] and another is a paper by Simone Conti and myself [**Conti and Karplus, 2019**].

The year 2019 is the 100th anniversary of the world's worst flu pandemic, which killed approximately 50 million people worldwide. Stimulated by this anniversary, the Gates Foundation made a call for proposals to develop a permanent flu vaccine. Victor, Simone and I submitted a proposal, which was guided by our experience with HIV. We were one of the eight groups that were selected for support and provided with a two year grant. (It is amusing to mention that we received notification from the Foundation in an email that first stated we were not selected followed by a statement that we were selected. Apparently the Foundation had originally thought one could have a flu vaccine candidate in the clinic within two years but then realized that was unrealistic.) Since our HIV research is completed (we are waiting for the results of Dennis Burton and his group), we have started on the flu vaccine project.

As can be seen from this section, I am still actively involved in research. The examples illustrate that I am trying to work in areas that are new to me, following the maxim cited earlier that one is more likely to do something original if one is not bound by a detailed knowledge of what has been done already. My hope is that in the future that is left to me, I will continue to be productive and have the satisfaction of discovering something new. My wife Marci comments at times that I still work 24/7, although in reality it is no longer the case, though my mind is often occupied with scientific problems, even if I am ostensibly doing something else.

Chapter 24

Life After the Nobel Prize

———————— ⚭ ————————

After receiving the Nobel Prize, I was invited to present keynote lectures at many more occasions than I could accept. I learned quickly that one has to turn down most invitations and accept only those where being a Nobel Laureate could have a positive effect. Many of the meetings were on subjects about which I knew little, certainly no more now than I did before receiving the Nobel Prize. An example is a meeting in Azerbaijan on economics with an honorarium of a $100,000, which I turned down. However, invitations to conferences concerned with climate change raised the possibility that I could be helpful in focusing the world on this vital subject. Since there were experts on climate change at Harvard (e.g., Michael McElroy, Steve Wofsy), I concluded it would be more constructive to write that they should be invited instead.

That people generally recognize the Nobel Prize as something very special was brought home to me one evening, not long after my Nobel Prize had been announced. My son Mischa and I were driving home from Milton, a town south of Boston, and it was getting dark. On the parkway near the Longwood Hospital area, Mischa swerved in such a way that one of the tires banged against the curb and blew out. There was not much traffic at the time, so we were able to park the car safely on a grassy area at the side of the road. On phoning 911, we were informed that a tow truck would shortly arrive. We waited in the car trying to keep warm because it was very cold. When the truck arrived, the driver came over to see how to maneuver so as to be able to tow the car. It being very much on my mind, I mentioned that I had just won a Nobel Prize. Immediately the driver said, "Go sit in the truck and stay warm and I will change the tire for you." Thus, rather than towing the car back to the garage, he did us the favor of taking care of the tire on the spot.

One of the first invitations I did accept was from the Beta Kappa Xi Honor Society and National Institute of Science, organized for students of historically black colleges, such as Lincoln College, Fisk University, and the University of the District of Columbia. These organizations have held annual meetings since the 1940s when the American Chemical Society (ACS) did not allow students from these colleges to attend meetings that were held in the South. The idea was to give the students an opportunity to hear exciting lectures and present their own work in posters or short oral presentations, essentially as they would have done at ACS annual meetings. Even though black students have been able to participate freely in ACS meetings for a long time, the honor society still exists and continues to hold an annual meeting. The one I attended was held in Houston, Texas in March 2014. When Kim Fenwick of the University of the District of Columbia, who was organizing the meeting, called to invite me, she clearly was not expecting that I would give a positive answer. After having looked into the background of the event, I accepted the invitation to be the "summa" lecturer. This was an opportunity to inspire young students to strive to go beyond their environment, to help give them the confidence to apply to Ivy League graduate schools, for example.

The meeting was well organized and had a series of talks, all delivered by outstanding scientists. Warren Washington, a Nobel Peace Prize recipient in 2007 as part of the Intergovernmental Panel on Climate Change (IPCC), gave an outstanding lecture on the dangers posed by climate change. I presented an updated version of the "Marsupial Lecture."[1] Although I attempt to expose the public to science in a way that is understandable, I am not always successful, as Marci has made clear. Nevertheless, I have introduced enough items of general interest to make the lecture enjoyable, even if not understandable, for everyone. Perhaps the most spectacular portion involves videos of dolphins with loud splashing sound effects. At the Houston meeting and subsequent such meetings, I urged the organizers to have the students read my 2006 "Spinach" article before the lecture. After the talk, I sat down on the stage and answered questions for nearly an hour. The questions ranged widely from those on the scientific aspects of my lecture to what it takes to get a Nobel Prize, to more personal ones about my early history and how I had felt when I first came to the United States. After the questions, many of the students asked me whether they could have a picture taken with me; some requested individual photos and others for the entire group from a given school (Figure 24.1).

---

[1] Before the Nobel Prize, I had rarely been invited to give a public lecture and I did not realize how useful the "Marsupial Lecture" would be. I first presented a version of it as part of the John Stouffer Lectureship at Stanford University in 2007, which requires you to present two lectures. One is a standard scientific lecture and the other a "public" lecture designed for a broader audience. As time goes on, I add general interest items that I come across to the "Marsupial Lecture" so that it continues to evolve.

Figure 24.1. With students at the Houston meeting after I presented my lecture

Almost everyone had a cell phone with a camera so that it was an efficient operation. As for me, I must admit I felt like a "rock star." I realize that for many of them, I was, in fact, a "rock star."

Another such invitation came from my niece, Beverly Hartline, the oldest daughter of my brother and his wife Betty. She is Vice Provost for Research at the University of Montana in Butte. I was invited to a convention to be held in February 2015 for middle school and high school students from Montana and the surrounding states. They were in Butte to participate in the Intermountain Junior Science and Humanities Fair and the Tech Regional High School Science Fair. I was to give the final address of the convention and I realized from my previous experience in Houston, that just my appearance, perhaps even more than the content of the lecture, could serve to make young students excited about their future in science. My presence (I cannot really say my lecture) had the same rock star effect as in Houston (Figure 24.2), and, of course, it proved to be a wonderful experience for me.

Andre Sali and other former students organized a celebration of my Nobel Prize and a precelebration of my 85th birthday in San Francicsco in October 2014. There were many lectures by former students. In addition there was a final lecture by David Chandler, who gave a wonderful overview of my work. It meant a great

Figure 24.2. With students after my lecture in Butte, Montana

deal to me because I had always admired his insightful approach to physical problems, though we had never published a joint paper; we had actually worked on a draft but never submitted it (Figure 24.3).

Perhaps the most unusual venue for the "Marsupial Lecture" was at the Universidad del Valle in Colombia in early November 2015. I was invited to come to Colombia, but was not inclined to travel there at the time. The person in charge, Professor Julio Cesar Arce Clavijo, suggested that I give my lecture from Harvard via WEBEX. This would make it possible for the students at their university to hear the lecture and see the slides. In addition, I would be able to see them and interact with them "as if I were there." Professor Clavijo persuaded me that it could be done effectively, on what is called the WEBEX system. After a certain amount of back-and-forth, testing WEBEX, and looking for a time that was possible for me and would also work for the students in Colombia, I decided to go ahead. Even though we had practiced everything beforehand, there were the inevitable complications. These technicalities were quickly resolved by an IT person, and I gave my lecture, including the sound effects of the dolphins. Then I invited questions. Both the professor and quite a few students had questions I could answer as if we were all together in the same lecture hall. Most rewarding

Figure 24.3. Karplusians and friends at the San Francisco celebration

was that Professor Clavijo sent me an email shortly afterward saying that several students were motivated to go on in science whereas before that they had planned to go on to business. What more can one ask?

The Malta VII Conference, which took place from November 15–20, 2015, in Rabat, the capital of Morocco, was a remarkable experience. These conferences are organized by Zafra Lerman with Malta chosen as the location of the first one because it is a neutral place where young scientists from all parts of the Middle East (e.g., Egypt, Lebanon, Palestine, Israel, and Syria) could meet and get to know each other. An important aspect of each conference is that a number of Nobel Prize winners are invited to present lectures. The goal is to foster the realization among the young people who attend that science can serve as a bridge to connect them. In fact, collaborations do develop, even if their countries are at war or do not have any other relations.

When I was invited by Zafra, I knew nothing about the Malta Conferences and initially I was somewhat skeptical. After an exchange of emails as well as discussions with Jean-Marie Lehn and Roald Hoffmann, both Nobel Laureates who had attended earlier conferences, I decided to accept. An element in my decision was the fact that the conference was being held in Morocco, which would give me a chance to visit some of the historic cities, such as Fez and Marrakesh, and hopefully add new photographs to my collection. In addition to local guides in the two towns, I had a driver, Abdellah, during the entire trip (see Chapter 17). Abdellah had strong opinions as to what would be of interest to a tourist, primarily buildings, but after a while he responded to my requests to see how people lived, where they shopped, and so on. Invariably when I asked him whether I could do something, his response was *insha' Allah*, freely translated as "God willing." Occasionally, we did get into trouble in the bazaars when I tried to photograph people, who, as it turned out, did not want to be photographed and started yelling at me. With my digital camera, it was much more difficult to use the method I had developed when I had my Leica with its Hector lens, so I mainly tried to get unposed pictures by taking them quickly (Figure 24.4).

Sometimes a shopkeeper would notice what I was doing and yell at me. Abdellah would then go over and talk with the person, and probably gave him a small amount of money. By then it was usually too late because the shopkeeper would pose with a big smile, the last thing I wanted.[2]

Figure 24.4. A shop in the Fez bazaar

[2]In Fez I also had a local guide, who took tourists around part-time to augment his income as a social worker. One day, while we were standing in front of the royal palace in Fez, he told me a story about King Mohammed VI, who came to power in 1999, after the death of his father. Mohammed VI is much more democratic than his father and is generally liked by the people. One day late at night, he visited a hospital and found that many of the nurses who should have been on duty were not there. When he asked why and was not given a satisfactory answer, rather than having them fired, he ordered the director to send the missing nurses to work for three years in small villages where there was a shortage of nurses. This seemed to me a very constructive way of dealing with the situation.

I have already described some examples of the opportunity the Nobel Prize has given me to inspire young people. The type of invitations I particularly like are those organized by students for students. Usually, they are organized by graduate students in chemistry, but sometimes even by undergraduates. In January 2018, the Harvard College Undergraduate Research Association organized a two-day meeting at Harvard as that year's venue of the National Collegiate Research Conferences. There were panels on various topics from entrepreneurship to how to get into medical school. Over 200 students from 81 universities participated. I was invited to be one of several keynote lecturers and presented the "Marsupial Lecture." It was gratifying to see the enthusiasm of the participants, though a significant fraction are likely to not have had the background to fully understand the lecture.

A different opportunity to participate in outreach is provided by the Nobel Laureates school visits, organized by Ed Shapiro, a Russian–American scientist and engineer. Their aim is to bring Nobel Laureates into high schools to inspire students to follow their interest in the sciences. The first one in which I participated was at the Cambridge Rindge and Latin School in Cambridge, Massachusetts in February 2017. Once the visit was arranged, I requested that the students watch my Nobel lecture and read my autobiography on the NobelPrize.org website. They then worked with their science teachers to develop a set of questions that avoided overlap. The program begins with a short introduction by Dr. Shapiro. Then the teachers and the student audience of about 100 or so juniors and seniors ask questions about anything, whether it concerns science, my life, or whatever else.

As a final vignette, I will mention what sometimes happens when I am riding in a taxi. One time, the driver turned out to have studied engineering in Nigeria. He was driving a taxi to earn money to support his extended family of 29 people: aunts and uncles, in addition to his parents, brothers, and sisters. At the same time, he was going to night school to earn a Master's degree in electrical engineering. The conversation turned to me and he asked me what I did. When I mentioned that I had won a Nobel Prize, he said what an honor it was for him to have me as a passenger. Similar reactions occur almost invariably when I mention my Nobel Prize. This reinforces my feeling that the prize has given me an opportunity to do some good in the world.

References

Abraham, M. J., Murtola, T., Schulz, R., Páll, S., Smith, J. C., Hess, B. and Lindahl, E. (2015) GROMACS: high performance molecular simulations through multi-level parallelism from laptops to supercomputers. *SoftwareX* **1–2**, pp. 19–25.

Abrahams, J. P., Leslie, A. G., Lutter, R. and Walker, J. E. (1994) Structure at 2.8 A resolution of F_1-ATPase from bovine heart mitochondria. *Nature* **370**, pp. 621–628.

Alberts, B. (1998) The cell as a collection of machines: preparing the next generation of molecular biologists. *Cell* **92**, pp. 291–294.

Albery, W. J. and Knowles, J. R. (1976) Evolution of enzyme function and the development of catalytic efficiency. *Biochemistry* **15**, pp. 5631–5640.

Alder, B. J. and Wainwright, T. E. (1957) Phase transition for a hard sphere system. *J. Chem. Phys.* **27**, pp. 1208–1209.

Anfinsen, C. B. (1973) Principles that govern folding of protein chains. *Science* **181**, pp. 223–230.

Artymiuk, P. J., Blake, C. C. F., Grace, D. E. P., Oatley, S. J., Phillips, D. C. and Sternberg, M. J. E. (1979) Crystallographic studies of the dynamic properties of lysozyme. *Nature* **280**, pp. 563–568.

Baldwin, R. L. (1990) Pieces of the folding puzzle. *Nature* **346**, pp. 409–410.

Baldwin, J. and Chothia, C. (1979) Haemoglobin: structural changes related to ligand binding and its allosteric mechanism. *J. Mol. Biol.* **129**, pp. 175–220.

Balint-Kurti, G. G. and Karplus, M. (1969) Multistructure valence-bond and atoms-in-molecules calculations for LiF, F_2, and F_2-. *J. Chem. Phys.* **50**, pp. 478–488.

Bash, P. A., Field, M. J., Davenport, R. C., Petsko, G. A., Ringe, D. and Karplus, M. (1991) Computer simulation and analysis of the reaction pathway of triosephosphate isomerase. *Biochemistry* **30**, pp. 5826–5832.

Bentinck-Smith, W. and Stouffer, E. (1991) *Harvard University History of Named Chairs*, Cambridge, Massachusetts.

Berendsen, H. (1976) Report of CECAM Workshop: models for protein dynamics. Orsay, May 24-July 17, 1976.

Beuhler, R. J., Bernstein, R. B. and Kramer, K. H. (1966) Observation of reactive asymmetry of methyl iodide crossed beam study of the reaction of rubidium with oriented methyl iodide molecules. *J. Am. Chem. Soc.* **88**, pp. 5331–5332.

Böckmann, R. A. and Grubmüller, H. (2002) Nanoseconds molecular dynamics simulation of primary mechanical energy transfer steps in FiATP synthase. *Nat. Struct. Biol.* **9**, pp. 198–202.

Bolhuis, P. G., Chandler, D., Dellago, C. and Geissler, P. L. (2002) Transition path sampling: throwing ropes over rough mountains passes in the dark. *Annu. Rev. Phys. Chem.* **53**, pp. 291–318.

Boresch, S. and Karplus, M. (1995) The meaning of component analysis: decomposition of the free energy in terms of specific interactions. *J. Mol. Biol.* **254**, pp. 801–807.

Boyer, P. D. (1993) The binding change mechanism for ATP synthase—some probabilities and possibilities. *Biochim. Biophys. Acta.* **1140**, pp. 215–250.

Bradshaw, W. H., Conrad, H. E., Corey, E. J., Gunsalus, I. C. and Lednicer, D. (1959) Microbiological degradation of (+)-camphor. *J. Am. Chem. Soc.* **81**, p. 5507.

Brooks, B. R., Brooks, C. L., III MacKerell, A. D., Jr., Nilsson, L., Petrella, R. J., Roux, B., Won, Y., Archontis, G., Bartels, C., Boresch, B., Caflisch, A., Caves, L., Cui, C., Dinner, A. R., Feig, M., Fischer, S., Gao, J., Hodoscek, M., Im, W., Kuczera, K., Lazaridis, T., Ma, J., Ovchinnikov, V., Paci, E., Pastor, R. W., Post, C. B., Pu, J. Z., Schaefer, M., Tidor, B., Venable, R. M., Woodcock, H. L., Wu, X., Yang, Y., York, D. M. and Karplus, M. (2009) CHARMM: the Biomolecular Simulation Program. *J. Comput. Chem.* **30**, pp. 1545–1614.

Brooks, B. R., Bruccoleri, R. E., Olafson, B. D., States, D. J., Swaminathan, S. and Karplus, M. (1983) CHARMM: a program for macromolecular energy, minimization, and dynamics calculations. *J. Comp. Chem.* **4**, pp. 187–217.

Brooks, B. R. and Karplus M. (1983) Harmonic dynamics of proteins: normal modes and fluctuations in bovine pancreatic trypsin inhibitor. *Proc. Natl. Acad. Sci. USA.* **80**, pp. 6571–6575.

Brooks, B. R. and Karplus M. (1985) Normal modes for specific motions of macro-molecules: application to the hinge-bending mode of lysozyme. *Proc. Natl. Acad. Sci. USA.* **82**, pp. 4995–4999.

Brooks, C. L., III, Karplus, M. and Pettitt, B. M. (1988) *Proteins: A Theoretical Perspective of Dynamics, Structure, and Thermodynamics.* New York: Wiley.

Brünger, A. T., Brooks, C. L, III. and Karplus, M. (1985) Active site dynamics of ribonu-clease. *Proc. Natl. Acad. Sci. USA.* **82**, pp. 8458–8462.

Brünger, A. T., Clore, G. M., Gronenborn, A. M. and Karplus, M. (1986) Three-dimensional structure of proteins determined by molecular dynamics with interproton distance restraints: application to crambin. *Proc. Natl. Acad. Sci. USA.* **83**, pp. 3801–3805.

Brünger, A. T., Huber, R. and Karplus, M. (1987) Trypsinogen-trypsin transition: a molecular dynamics study of induced conformational change in the activation domain. *Biochemistry* **26**, pp. 5153–5162.

Brünger, A. T. and Karplus, M. (1991) Molecular dynamics simulations with experimental restraints. *Acc. Chem. Res.* **24**, pp. 54–61.

Brünger, A. T., Kuriyan, J. and Karplus, M. (1987) Crystallographic R factor refinement by molecular dynamics. *Science* **235**, pp. 458–460.

Bryngelson, J. D. and Wolynes, P. G. (1989) Intermediates and barrier crossing in a random energy model (with applications to protein folding). *J. Phys. Chem.* **93**, pp. 6902–6915.

Camacho, C. J. and Thirumalai, D. (1993) Kinetics and thermodynamics in model proteins. *Proc. Natl. Acad. Sci. USA.* **90**, pp. 6369–6372.

Cammarata, M., Levantino, M., Wulff, M. and Cupane, A. (2010) Unveiling the timescale of the R-T transition in human hemoglobin. *J. Mol. Biol.* **400**, pp. 951–962.

Case, D. A. and Karplus, M. (1979) Dynamics of ligand binding to heme proteins. *J. Mol. Biol.* **132**, pp. 343–368.

Caves, T. C. and Karplus, M. (1969) Perturbed Hartee-Fock theory. I. Diagrammatic double-perturbation analysis. *J. Chem. Phys.* **50**, pp. 3649–3661.

Chan, H. S. (1995) Kinetics of protein folding. *Nature* **373**, pp. 664–665.

Changeux, J.-P. and Edestein, S. J. (2005) *Science* **308**, pp. 1424–1428.

Colonna-Cesari, F., Perahia, D., Karplus, M., Ecklund, H., Brändén, C. I. and Tapia, O. (1986) Interdomain motion in liver alcohol dehydrogenase: structural and energetic analysis of the hinge bending mode. *J. Biol. Chem.* **261**, pp. 15273–15280.

Conroy, H. (1960) Nuclear magnetic resonance in organic structural elucidation. *Adv. Org. Chem.* **2**, pp. 265–294.

Conti S. and Karplus, M. (2019) Estimation of the breadth of CD4bs targeting HIV antibodies by molecular modeling and machine learning. *PLOS Comput. Biol.* **15**, e1006954.

Cui, Q. and Karplus, M. (2002) Quantum mechanical/molecular mechanical studies of the triosephosphate isomerase-catalyzed reactions: verification of methodology and analysis of the reaction mechanisms. *J. Phys. Chem. B* **106**, pp. 1768–1798.

Cui, Q. and Karplus, M. (2003) Catalysis and specificity in enzymes: a study of triosephosphate isomerase and comparison with methyl glyoxal synthase. *Adv. Protein Chem.* **66**, pp. 315–372.

Cui, Q. and Karplus, M. (2008) Allostery and cooperativity revisited. *Protein Sci.* **17**, pp. 1295–1307.

Cusack, S., Smith, J., Finney, J., Karplus, M. and Trewhella, J. (1986) Low frequency dynamics of proteins studied by neutron time-of-flight spectroscopy. *Physica* **136B**, pp. 256–259.

Dalton, L. (2003) Karplus equation. *Chem. Eng. News* **81**, pp. 37–39.

Deisenhofer, J. and Steigemann, W. (1975) Crystallographic refinement and the structure of the bovine pancreatic trypsin inhibitor at 1.5 Å resolution. *Acta Crystallogr. B* **31**, pp. 238–250.

Dill, K. A., Bromberg, S., Yue, K., Fiebig, K. M., Yee, D. P., Thomas, P. D. and Chan, H. S. (1995) Principles of protein folding – a perspective from simple exact models. *Protein Sci.* **4**, pp. 561–602.

Dinner, A. R., Blackburn, G. M. and Karplus, M. (2001) Uracil-DNA glycosylase acts by substrate autocatalysis. *Nature* **413**, pp. 752–755.

Dobson, C. M. (2003) Protein folding and misfolding. *Nature* **426**, pp. 884–890.

Dobson, C. M. and Karplus, M. (1986) Internal motion of proteins: nuclear magnetic resonance measurements and dynamic simulations. *Methods Enzymol.* **131**, pp. 362–389.

Dobson, C. M., Sali A. and Karplus M. (1998) Protein folding: a perspective from theory and experiment. *Angew. Chem. Int. Ed.* **37**, pp. 868–893.

Eaton, W. A., Henry, E. R., Hofrichter, J. and Mozzarelli, A. (1999) Is cooperative oxygen binding by hemoglobin really understood? *Nature Struct. Biol.* **6**, pp 351–358.

Elber, R. and Karplus, M. (1990) Enhanced sampling in molecular dynamics: use of the time-dependent Hartree approximation for a simulation of carbon monoxide diffusion through myoglobin. *J. Am. Chem. Soc.* **112**, pp. 9161–9175.

Farkas, A. and Farkas, L. (1935) Experiments on heavy hydrogen. V. The elementary reactions of light and heavy hydrogen. The thermal conversion of ortho-deuterium and the interaction of hydrogen and deuterium. *Proc. R. Soc. Lond. A* **152**, pp. 124–151.

Feierberg, I. and Aqvist, J. (2002) Computatational modeling of enzymatic keto-enol isomerization reactions. *Theor. Chem. Acc.* **108**, pp. 71–84.

Feynman, R. P., Leighton, R. B. and Sands, M. (1963) *The Feynman Lectures in Physics.* Addison-Wesley.

Field, M. J., Bash, P. A. and Karplus, M. (1990) A combined quantum mechanical and molecular mechanical potential for molecular dynamics simulations. *J. Comp. Chem.* **11**, pp. 700–733.

Fischer, S., Olsen, K. W., Nam, K. and Karplus, M. (2011) Unsuspected pathway of the allosteric transition in hemoglobin. *Proc. Natl. Acad. Sci. USA* **108**, pp. 5608–5613.

Frauenfelder, H., Hartmann, H., Karplus, M., Kuntz, I. D., Jr., Kuriyan, J., Parak, F., Petsko, G. A., Ringe, D., Tilton, R. F., Jr., Connolly, M. L. and Max, M. (1987) Thermal expansion of a protein. *Biochemistry* **26**, pp. 254–261.

Frauenfelder, H., Petsko, G. A. and Tsernoglou, D. (1979) Temperature-dependent x-ray diffraction as a probe of protein structural dynamics. *Nature* **280**, pp. 558–563.

Freeman, D. L. and Karplus, M. (1976) Many-body perturbation theory applied to molecules: analysis and correlation energy calculation for Li_2, F_2 and H_3. *J. Chem. Phys.* **64**, pp. 2641–2659.

Friedman, R. (2001) *The Politics of Excellence: Behind the Nobel Prize in Science*, New York, Henry Holt and Co. Chapter 7.

Gao, J., Kuczera, K., Tidor, B. and Karplus, M. (1989) Hidden thermodynamics of mutant proteins: a molecular dynamics analysis. *Science* **244**, pp. 1069–1072.

Gao, Y. Q., Yang, W. and Karplus, M. (2005) A structure-based model for synthesis and hydrolysis of ATP by F_1-ATPase. *Cell* **123**, pp. 195–205.

Gao, Y. Q., Yang, W., Marcus, R. A. and Karplus, M. (2003) A model for the cooperative free energy transduction and kinetics of ATP hydrolysis by F_1-ATPase. *Proc. Natl. Acad. Sci. USA* **100**, pp. 11339–11344.

Garcia-Viloca, M., Gao, J., Karplus, M. and Truhlar, D. G. (2004) How enzymes work: aanalysis by modern rate theory and computer simulations. *Science* **303**, pp. 186–195.

Gelin, B. R. (1976) Application of empirical energy functions to conformational problems in biochemical systems. PhD thesis. Harvard University.

Gelin, B. R. and Karplus, M. (1975) Sidechain torsional potentials and motion of amino acids in proteins: bovine pancreatic trypsin inhibitor. *Proc. Natl. Acad. Sci. USA* **72**, pp. 2002–2006.

Gelin, B. R. and Karplus, M. (1977) Mechanism of tertiary structural change in hemoglobin. *Proc. Natl. Acad. Sci. USA* **74**, pp. 801–805.

Gilardi, R., Karle, I. L., Karle, J. and Sperling, W. (1971) Crystal structure of visual chromophores, 11-cis and all-trans retinal. *Nature* **232**, pp. 187–189.

Go, N. (1983) Theoretical studies of protein folding. *Annu. Rev. Biophys. Bioeng.* **12**, pp. 183–210.

Godfrey, M. and Karplus, M. (1968) Theoretical investigation of reactive collisions in molecular beams: K+Br$_2$. *J. Chem. Phys.* **49**, pp. 3602–3609.

Goldstern Family Tree. Based on "The Goldstern Family Story" by Norbert Goldstern in the Goldstern Family Tree, compiled by Walter Goldstern.

Griscom, L. (1949) *The Birds of Concord*, Cambridge, MA, Harvard University Press.

Harrison, S. C. and Durbin, R. (1985) Is there a single pathway for the folding of a polypeptide chain? *Proc. Natl. Acad. Sci. USA* **82**, pp. 4028–4030.

Harvey, S. C., Prabhakaran, M., Mao, B. and McCammon, J. A. (1984) Phenylalanine transfer RNA: molecular dynamics simulation. *Science* **223**, pp. 1189–1191.

Hemley, R. J., Dinur, U., Vaida, V. and Karplus, M. (1985) Theoretical study of the ground and excited singlet states of styrene. *J. Am. Chem. Soc.* **107**, pp. 836–844.

Henzler-Wildman, K. A., Thai, C., Lei, M., Ott, M., Wolf-Watz, M., Fenn, T., Pozharski, W., Wilson, M. A., Petsko, G. A., Karplus, M., Hüner, C. G. and Kern, D. (2007) Intrinsic motions along an enzymatic reaction trajectory. *Nature* **450**, pp. 838–844.

Hirschfelder, J. A., Eyring, H. and Topley, B. (1936) Reactions involving hydrogen molecules and atoms. *J. Chem. Phys.* **4**, pp. 170–177.

Honig, B., Hudson, B., Sykes, B. D. and Karplus, M. (1971) Ring orientation in β-ionone and retinals. *Proc. Natl. Acad. Sci. USA* **68**, pp. 1289–1293.

Honig, B. and Karplus, M. (1971) Implications of torsional potential of retinal isomers for visual excitation. *Nature* **229**, pp. 558–560.

Honig, B., Warshel, A. and Karplus, M. (1975) Theoretical studies of the visual chromophore. *Acc. Chem. Res.* **8**, pp. 92–100.

Huszar, D., Theoclitou, M. E., Skolnik, J. and Herbst, R. (2009) Kinesin motor proteins as targets for cancer therapy. *Cancer Metastasis Rev.* **28**, pp. 197–208.

Hvidt, A. and Nielsen, S. O. (1966) Hydrogen exchange in proteins. *Adv. Protein Chem.* **21**, pp. 287–286.

Hwang, W., Lang, M. J. and Karplus, M. (2017) Kinesin motility is driven by subdomain dynamics. *eLife* **6**, e28948.

Ichiye, T. and Karplus, M. (1983) Fluorescence depolarization of tryptophan residues in proteins: a molecular dynamics study. *Biochemistry* **22**, pp. 2884–2893.

Imai, K. and Osawa, E. (1989) An extension of multiparameteric Karplus equation. *Tetrahedron Lett.* **30**, pp. 4251–4254.

Irikura, K. K., Tidor, B., Brooks, B. R. and Karplus, M. (1985) Transition from B to Z DNA: contribution of internal fluctuations to the configurational entropy difference. *Science* **229**, pp. 571–572.

Islam, S. A., Karplus, M. and Weaver, D. L. (2002) Application of the diffusion-collision model to the folding of three-helix bundle proteins. *J. Mol. Biol.* **318**, pp. 199–215.

Islam, S. A., Karplus, M. and Weaver, D. L. (2004) The role of sequence and structure in protein folding kinetics: the diffusion-collision model applied to proteins L and G. *Structure* **12**, pp. 1833–1845.

Jiang, Y. L., Ichikawa, Y., Song, F. and Stivers, J. T. (2003) Powering DNA repair through substrate electrostatic interactions. *Biochemistry* **42**, pp. 1922–1929.

Joseph, D., Petsko, G. A. and Karplus, M. (1990) Anatomy of a protein conformational change: hinged "lid" motion of the triosephosphate isomerase loop. *Science* **249**, pp. 1425–1428.

Karplus, M. (1952) Bird activity in the continuous daylight of arctic summer. *Ecology* **33**, pp. 129–134.

Karplus, M. (1956) Charge distribution in the hydrogen molecule. *J. Chem. Phys.* **25**, pp. 605–606.

Karplus, M. (1959a) Contact electron-spin interactions of nuclear magnetic moments. *J. Chem. Phys.* **30**, pp. 11–15.

Karplus, M. (1959b) Interpretation of the electron-spin resonance spectrum of the methyl radical. *J. Chem. Phys.* **30**, pp. 15–18.

Karplus, M. (1960a) Theory of proton coupling constants in unsaturated molecules. *J. Am. Chem. Soc.* **82**, p. 4431.

Karplus, M. (1960b) Weak interactions in molecular quantum mechanics. *Rev. Mod. Phys.* **32**, pp. 455–460.

Karplus, M. (1963) Vicinal proton coupling in nuclear magnetic resonance. *J. Am. Chem. Soc.* **85**, p. 2870.

Karplus, M. (1968) Structural implications of reaction kinetics, in *Structural Chemistry and Molecular Biology*, eds. A. Rich, and N. Davidson, San Francisco, Freeman, pp. 837–847.

Karplus, M. (1982) Dynamics of proteins. *Ber. Bunsen-Ges. Phys. Chem.* **86**, pp. 386–395.

Karplus, M. (1996) Theory of vicinal coupling constants, in *Encyclopedia of Nuclear Magnetic Resonance. Vol. 1: Historical Perspectives*, ed. D. M. Grant, R. K. Harris, New York, Wiley, pp. 420–422.

Karplus, M. (1997) The Levinthal paradox: yesterday and today. *Fold. Des.* **2**, pp. 569–576.

Karplus, M. (2002) Molecular dynamics simulations of biomolecules. *Acc. Chem. Res.* **35**, pp. 321–323.

Karplus, M. (2011) *Images from the 50's* (Blurb.com).

Karplus, M. (2014a) Development of multiscale models for complex chemical systems: from $H+H_2$ to biomolecules (Nobel Lecture). *Angew. Chem. Int. Ed.* **53**, pp. 9992–10005.

Karplus, M. (2014b) The Nobel Prizes 2013, in *Published on behalf of The Nobel Foundation*, ed. K. Grandin, USA, Science History Publications, pp. 41–96.

Karplus, M. (2016) Preface, Special Issue: Free Energy Simulations. *Mol. Simulat.* **42**, pp. 1044–1045.

Karplus, M. and Fraenkel, G. K. (1961) Theoretical interpretation of carbon-13 hyperfine interactions in electron spin resonance spectra. *J. Chem. Phys.* **35**, pp. 1312–1323.

Karplus, M. and Godfrey, M. (1966) Quasiclassical trajectory analysis for the reaction of potassium atoms with oriented methyl iodide molecules. *J. Am. Chem. Soc.* **88**, p. 5332.

Karplus, S. and Karplus, M. (1972) Nuclear magnetic resonance determination of the angle ψ in peptides. *Proc. Natl. Acad. Sci. USA* **69**, pp. 3204–3206.

Karplus, R. and Kroll, N. M. (1950) Fourth-order corrections in quantum electrodynamics and the magnetic moment of the electron. *Phys. Rev.* **77**, pp. 536–549.

Karplus, M., Kuppermann, A. and Isaacson, L. M. (1958) Quantum-mechanical calculation of one-electron properties. I. General formulation. *J. Chem. Phys.* **29**, pp. 1240–1246.

Karplus, M. and Kuriyan, J. (2005) Molecular dynamics and protein function. *Proc. Natl. Acad. Sci. USA* **102**, pp. 6679–6685.

Karplus, M. and Kushick, J. N. (1981) Method for estimating the configurational entropy of macromolecules. *Macromolecules* **14**, pp. 325–332.

Karplus, M. and Lavery, R. (2014) Significance of molecular dynamics simulations for life sciences. *Isr. J. Chem.* **54**, pp. 1042–1051.

Karplus, M., Lawler, R. G. and Fraenkel, G. K. (1965) Electron spin resonance studies of deuterium isotope effects. A novel resonance-integral perturbation. *J. Am. Chem. Soc.* **87**, p. 5260.

Karplus, M. and McCammon, J. A. (2002) Molecular dynamics simulations of biomolecules. *Nat. Struct. Biol.* **9**, pp. 646–652.

Karplus, M. and Porter, R. N. (1970) *Atoms and Molecules: An Introduction for Students of Physical Chemistry*, Menlo Park, CA, Benjamin Cummins.

Karplus, M., Porter, R. N. and Sharma, R. D. (1965) Exchange reactions with activation energy. I. Simple barrier potential for (H, H_2). *J. Chem. Phys.* **43**, pp. 3259–3287.

Karplus, M., Sali, A. and Shakhnovich, E. (1995) Comment: kinetics of protein folding. *Nature* **373**, p. 665.

Karplus, M. and Weaver, D. L. (1976) Protein-folding dynamics. *Nature* **260**, pp. 404–406.

Karplus, M. and Weaver, D. L. (1994) Protein folding dynamics: The diffusion-collision model and experimental data. *Protein Sci.* **3**, pp. 650–668.

Kern, D. and Zuiderweg, E. R. (2003) The role of dynamics in allosteric regulation. *Curr. Opin. Struc. Biology* **13**, pp. 748–757.

Kinosita, K., Jr., Adachi, K. and Itoh, H. (2004) Rotation of F_1-ATPase: how an ATP-driven molecular machine works. *Annu. Rev. Biophys. Biomol. Struct.* **33**, pp. 245–268.

Kirkwood, J. G. and Oppenheim, I. (1961) *Chemical Thermodynamics*, New York, McGraw Hill.

Klein, M., Krivov, S. V., Ferrer, A. J., Luo, L., Samuel, A. D. T. and Karplus, M. (2017) Exploratory search during directed navigation in *C. elegans* and *Drosophila* larva. *eLife* **6**, pp. e30503.

Kline, A. D., Braun, W. and Wüthrich, K. (1988) Determination of the complete 3-dimensional structure of the alpha-amylase inhibitor tendamistat in aqueous-solution by nuclear magnetic resonance and distance geometry. *J. Mol. Biol.* **204**, pp. 675–724.

Kuppermann, A., Karplus, M. and Isaacson, L. M. (1959) The quantum-mechanical calculation of one-electron properties. II. One-and two-center moment integrals. *Zeit. Nat.* **14a**, pp. 311–318.

Kuriyan, J., Konforti, B. and Wemmer, D. (2017) *The Molecules of Life*, New York, Taylor & Francis.

Kuriyan, J. and Weis, W. I. (1991) Rigid protein motion as a model for crystallographic temperature factors. *Proc. Natl. Acad. Sci. USA* **88**, pp. 2773–2777.

Lawler, R. G., Bolton, J. R., Karplus, M. and Fraenkel, G. K. (1967) Deuterium isotope effects in the electron spin resonance spectra of naphthalene negative ions. *J. Chem. Phys.* **47**, pp. 2149–2165.

Lee, A. W., Karplus, M., Poyart, C. and Bursaux, E. (1988) Analysis of proton release in oxygen binding by hemoglobin: implications for the cooperative mechanism. *Biochemistry* **27**, pp. 1285–1301.

Levitt, M. and Lifson, S. (1969) Refinement of protein conformations using a macromolecular energy minimization procedure. *J. Mol. Biol.* **46**, pp. 269–279.

Levy, R. M., Karplus, M. and Wolynes, P. G. (1981) NMR relaxation parameters in molecules with internal motion: exact Langevin trajectory results compared with simplified relaxation models. *J. Am. Chem. Soc.* **103**, pp. 5998–6011.

Lifson, S. and Warshel, A. (1969) Consistent force field for calculations of conformations, vibrational spectra, and enthalpies of cycloalkanes and n-alkane molecules. *J. Chem. Phys.* **49**, pp. 5116–5129.

Linderstrom-Lang, K. (1955) Deuterium exchange between peptides and water. *Chem. Soc. Spec. Publ.* **2**, p. 1.

Lindorff-Larsen, K., Piana, S., Dror, R. O. and Shaw, D. E. (2011) How fast-folding proteins fold. *Science* **334**, pp. 517–520.

Liu, B. (1973) Ab-initio potential-energy surface for linear H_3. *J. Chem. Phys.* **58**, pp. 1925–1937.

Lodi, P. J. and Knowles, J. R. (1991) Neutral imidazole is the electrophile in the reaction catalyzed by triosephosphate isomerase: structural origins and catalytic implications, *Biochemistry* **30**, pp. 6948–6969.

Ma, J., Flynn, T. C., Cui, Q., Leslie, A. G. W., Walker, J. E. and Karplus M. (2002) A dynamic analysis of the rotation mechanism for conformational change in F_1-ATPase. *Structure* **10**, pp. 921–931.

Ma, A., Hu, J., Karplus, M. and Dinner, A. R. (2006) Implications of alternative substrate binding modes for catalysis by Uracil-DNA glycosylase: an apparent discrepancy resolved. *Biochemistry* **45**, pp. 13687–13696.

Ma, J., Sigler, P. B., Xu, Z. and Karplus, M. (2000) A dynamic model for the allosteric mechanism of GroEL. *J. Mol. Biol.* **302**, pp. 303–313.

Mao, H. Z. and Weber, J. (2007) Identification of the β_{TP} site in the x-ray structure of F_1-ATPase as the high-affinity catalytic site. *Proc. Natl. Acad. Sci. U.S.A* **104**, pp. 18478–18483.

Mark, A. E. and van Gunsteren, W. F. (1994) Decomposition of the free energy of a system in terms of specific interactions: implications for theoretical and experimental studies. *J. Mol. Biol.* **240**, pp. 167–176.

McCammon, J. A., Gelin, B. R. and Karplus, M. (1977) Dynamics of folded proteins. *Nature* **267**, pp. 585–590.

Mierke, D. F., Huber, T. and Kessler, H. (1994) Coupling-constants again: experimental restraints in structure refinement. *J. Comp. Aided Mol. Des.* **8**, pp. 29–40.

Milgram, S. (1967) The small world problem. *Psychol. Today* **1**, pp. 61–67.

Milgram, S. and Travers, J. (1969) An experimental study of the small world problem. *Sociometry* **32**, pp. 425–443.

Miranker, A. and Karplus, M. (1991) Functionality maps of binding sites: a multiple copy simultaneous search method. *Proteins Struct. Funct. Genet.* **11**, pp. 29–34.

Moffitt, W. (1954) Atomic valence states and chemical binding. *Rep. Prog. Phys.* **17**, pp. 173–200.

Monod, J., Wyman, J. and Changeux, J. P. (1965) On the nature of allosteric transitions: a plausible model. *J. Mol. Biol.* **12**, pp. 88–118.

Moore, G. F. (1965) Cramming more components into integrated circuits. *Electronics* **38**, pp. 114–117.

Morokuma, K., Eu, B. C. and Karplus, M. (1969) Collision dynamics and the statistical theories of chemical reactions. I. Average cross section from transition-state theory. *J. Chem. Phys.* **51**, pp. 5193–5203.

Morokuma, K. and Karplus, M. (1971) Collision dynamics and the statistical theories of chemical reactions. II. Comparison of reaction probabilities. *J. Chem. Phys.* **55**, pp. 63–75.

Nadler, W., Brünger, A. T., Schulten, K. and Karplus, M. (1987) Molecular and stochastic dynamics of proteins. *Proc. Natl. Acad. Sci. USA* **84**, pp. 7933–7937.

Nam, K. and Karplus, M. (2019) Insight into the origin of the high energy-conversion efficiency of F_1-ATPase. *Proc. Natl. Acad. Sci. USA* **116**, pp. 15924–15929.

Nam, K., Pu, J. and Karplus, M. (2014) Trapping the ATP binding state leads to a detailed understanding of the F_1-ATPase mechanism. *Proc. Natl. Acad. Sci. USA* **50**, pp. 17851–17856.

Nam, K., Verdine, G. L. and Karplus, M. (2009) Analysis of an anomalous mutant of MutM DNA glycosylase leads to new insights into the catalytic mechanism. *J. Am. Chem. Soc.* **131**, pp. 18208–18209.

Nilsson, L., Clore, G. M., Gronenborn, A. M., Brünger, A. T. and Karplus, M. (1986) Structure refinement of oligonucleotides by molecular dynamics with nuclear Overhauser effect interproton distance restraints: application to 5'd(C-G-T-A-C-G)$_2$. *J. Mol. Biol.* **188**, pp. 455–475.

Noji, H., Yasuda, R., Yoshida, M. and Kinosita, K., Jr. (1997) Direct observation of the rotation of F_1-ATPase. *Nature* **386**, pp. 299–302.

Olejniczak, E. T., Dobson, C. M., Levy, R. M. and Karplus, M. (1984) Motional averaging of proton nuclear Overhauser effects in proteins. Predictions from a molecular dynamics simulation of lysozyme. *J. Am. Chem. Soc.* **106**, pp. 1923–1930.

Ottenbacher, A. (1999) *Eugenie Goldstern*, Wien, Mandelbaum Verlag.

Ovchinnikov, V. and Karplus, M. (2014) Investigations of α-helix $\rightarrow$ β-sheet transition pathways in a miniprotein using the finite-temperature string method. *J. Chem. Phys.* **140**, pp. 175103.1–175103.18.

Ovchinnikov, V., Louveau, J. E., Barton, J. P., Karplus, M. and Chakraborty, A. K. (2018) Role of framework mutations and antibody flexibility in the evolution of broadly neutralizing antibodies. *eLife* **6**, p. e33038.

Papanek, E. and Linn, E. (1975) *Out of the Fire*, New York, William Morrow & MoCo.

Pauling, L. (1946) Molecular architecture and biological reactions. *Chem. Eng. News* **24**, pp. 1375–1377.

Perutz, M. (1971) Stereochemistry of cooperative effects in haemoglobin. *Nature* **232**, pp. 408–413.

Phillips, D. C. (1981) Closing remarks, in *Biomolecular Stereodynamics*, Vol 2, ed. R. H. Sarma, Guilderland, Adenine Press, pp. 497–498.

Phillips, J. C., Braun, R., Wang, W., Gumbart, J., Tajkhorshid, E., Villa, E., Chipot, C., Skeel, R. D., Kalé, L. and Schulten, K. (2005) Scalable molecular dynamics with NAMD. *Comp. Chem.* **26**, pp. 1781–1802.

Porter, R. N. and Karplus M. (1964) Potential energy surface for H_3. *J. Chem. Phys.* **40**, pp. 1105–1115.

Post, C. B. and Dobson, C. M. (2005) Frontiers in computational biophysics: a symposium in honor of Martin Karplus. *Structure* **13**, pp. 949–952.

Provence Beyond website, Michelin Guide History (www.beyond.fr) Last accessed 29 April 2020.

Purins, D. and Karplus M. (1969) Spin delocalization and vibrational-electronic interaction in the toluene ion-radicals. *J. Chem. Phys.* **50**, pp. 214–233.

Qi, Y., Spong, M. C., Nam, K., Banerjee, A., Jiralerspong, S., Karplus, M. and Verdine, G. L. (2009) Encounter and extrusion of an intrahelical lesion by a DNA repair enzyme. *Nature* **462**, pp. 762–769.

Rahman, A. (1964) Correlations in motion of atoms in liquid argon. *Phys. Rev.* **136**, pp. A405–A411.

Sali, A., Shakhnovich, E. and Karplus M. (1994) How does a protein fold? *Nature* **369**, pp. 248–251.

Schatz, G. C. (2000) Perspective on "Exchange reactions with activation energy. I. Simple barrier potential for (H, H_2)." *Theor. Chem. Acc.* **103**, pp. 270–272.

Schatz, G. C. and Kuppermann, A. (1977) Quantum-mechanical reactive scattering: accurate 3-dimensional calculation. *J. Chem. Phys.* **65**, pp. 668–692.

Schekkerman, H., Tulp, I., Piersma, T. and Visser, G. H. (2003) Mechanisms promoting higher growth rate in arctic than in temperate shorebirds. *Ecophysiology* **134**, pp. 332–342.

Scheraga, H. A. (1968) Calculations of the conformations of small molecules. *Adv. Phys. Org. Chem.* **6**, pp. 103–184.

Schowen, R. L. (1978) Catalytic power and transition-state stabilization, in *Transition States of Biochemical Processes*, ed. R. D. Gandour. and R. L. Schowen, New York, Plenum, pp. 77–114.

Schroedinger, E. (1944) *What is Life?* Cambridge, Cambridge University Press.

Schulten, K. and Karplus, M. (1972) On the origin of a low-lying forbidden transition in polyenes and related molecules. *Chem. Phys. Lett.* **14**, pp. 305–309.

Shavitt, I. and Karplus, M. (1962) Multicenter integrals in molecular quantum mechanics. *J. Chem. Phys.* **36**, pp. 550–551.

Shavitt, I., Stevens, R. M., Minn, F. L. and Karplus, M. (1968) Potential-energy surface for H_3. *J. Chem. Phys.* **48**, pp. 2700–2713.

Shaw, D. E., Maragakis, P., Lindorff-Larsen, K., Piana, S., Dror, R. O., Eastwood, M. F., Bank, J. A., Jumper, J. M., Salmon, J. K., Shan, Y. and Wriggers, W. (2010) Atomic-level characterization of the structural dynamics of proteins. *Science* **330**, pp. 341–346.

Shulman, R. G., Glarum, S. H. and Karplus M. (1971) Electronic structure of cyanide complexes of hemes and heme proteins. *J. Mol. Biol.* **57**, pp. 93–115.

Simonson, T., Archontis, G. and Karplus, M. (2002) Free energy simulations come of age: protein-ligand recognition. *Acc. Chem. Res.* **35**, pp. 430–437.

Smith, J., Cusack, S., Pezzeca, U., Brooks, B. R. and Karplus, M. (1986) Inelastic neutron scattering analysis of low frequency motion in proteins: a normal mode study of the bovine pancreatic trypsin inhibitor. *J. Chem. Phys.* **85**, pp. 3636–3654.

Spiro, T. G. and Balakrishnan, G. (2010) Quaternary speeding in hemoglobin. *J. Mol. Biol.* **400**, pp. 949–950.

Stillinger, F. H. and Rahman, A. (1974) Improved simulation of liquid water by molecular-dynamics. *J. Chem. Phys.* **60**, pp. 1545–1557.

Szabo. A. and Karplus, M. (1972) A mathematical model for structure-function relations in hemoglobin. *J. Mol. Biol.* **72**, pp. 163–197.

Szent-Györgyi, A. (1948) *Nature of Life, A Study of Muscle*, New York, Academic Press.

Taly, A., Corringer, P. J., Grutter, T., Prado de Carvalho, L., Karplus, M., and Changeux, J. P. (2006) Implications of the quaternary twist allosteric model for the physiology and pathology of nicotinic acetylcholine receptors. *Proc. Natl. Acad. Sci. USA* **103**, pp. 16965–16970.

Taylor, E. H. and Datz, S. (1955) Study of chemical reaction mechanisms with molecular beams: the reaction of K with HBr. *J. Chem. Phys.* **23**, pp. 1711–118.

Truhlar, D. G. and Wyatt, R. E. (1976) History of H_3 kinetics. *Ann. Rev. Phys Chem.* **27**, pp. 1–43.

van der Vaart, A., Ma, J., and Karplus, M. (2004) The unfolding action of GroEL on a protein substrate. *Biophys. J.* **87**, pp. 562–573.

Vendruscolo, M. and Dobson, C. M. (2010) Protein dynamics: Moore's law in molecular biology. *Curr. Biol.* **21**, pp. R68–R70.

Vendruscolo, M., Paci. E., Dobson, C. M. and Karplus, M. (2001) Three key residues form a critical contact network in a protein folding transition state. *Nature* **409**, pp. 641–645.

Wall, F. T. and Porter, R. N. (1963) Sensitivity of exchange-reaction probabilities to the potential-energy surface. *J. Chem. Phys.* **39**, pp. 3112–3117.

Warshel, A. and Karplus, M. (1974) Calculation of $\pi\pi^*$ excited state conformations and vibronic structure of retinal and related molecules. *J. Am. Chem. Soc.* **96**, pp. 5677–5689.

Weber, J., Wilke-Mounts, S., Lee, R. S., Grell, E. and Senior, A. E. (1993) Specific placement of tryptophan in the catalytic sites of Escherichia coli F_1-ATPase provides a direct probe of nucleotide binding: maximal hydrolysis occurs with all three sites occupied. *J. Biol. Chem.* **268**, pp. 20126–20133.

Wolfenden, R. and Snider, M. J. (2001) The depth of chemical time and the power of enzymes as catalysts. *Acc. Chem. Res.* **34**, pp. 938–945.

Wolynes, P. G. (2005) Energy landscapes and solved protein-folding problems. *Phil. Trans. R. Soc. A* **363**, pp. 453–454.

Wong, C. F. and McCammon, J. A. (1986) Dynamics and design of enzymes and inhibitors. *J. Am. Chem. Soc.* **108**, pp. 3830–3832.

Wyatt, R. and Elkowitz, A. (1975) Quantum mechanical reaction cross sections for the three-dimensional hydrogen exchange reaction. *J. Chem. Phys.* **62**, pp. 2504–2506.

Yang, W., Bitetti-Putzer, R. and Karplus, M. (2004) Free energy simulations: use of reverse cumulative averaging to determine the equilibrated region and the time required for convergence. *J. Chem. Phys.* **120**, pp. 2618–2628.

Yang, W., Gao, Y. Q., Cui, Q., Ma, J. and Karplus, M. (2003) The missing link between thermodynamics and structure in F_1-ATPase. *Proc. Natl. Acad. Sci. USA* **100**, pp. 874–879.

Young, M. A., Gonfloni, S., Superti-Furga, G., Roux, B. and Kuriyan, J. (2001) Dynamic coupling between the SH_2 and SH_3 domains of c-Src and hck underlies their inactivation by C-terminal tyrosin phosphorylation. *Cell* **105**, pp. 115–126.

Zheng, G., Schaefer, M. and Karplus, M. (2013) Hemoglobin Bohr effect: atomic origin of the histidine residue contributions. *Biochemistry* **52**, pp. 8539–8555.

Zhou, Y. and Karplus, M. (1999) Interpreting the folding kinetics of a helical protein. *Nature* **401**, pp. 400–403.

Zhou, Y., Zhou, H. and Karplus, M. (2003) Cooperativity in *scapharca* dimeric Hemoglobin: simulation of binding intermediates and elucidation of the role of interfacial water. *J. Mol. Biol.* **326**, pp. 593–606.

Karplusians: 1955–2019

Ivana Adamovic

Yuri Alexeev

David H. Anderson

Ioan Andricioaei

Yasuhide Arata

Georgios Archontis

Gabriel G. Balint-Kurti

Christian Bartels

Paul Bash

Donald Bashford

Mark Bathe

Oren M. Becker

Robert Best

Anton Beyer

Robert Birge

Ryan Bitetti-Putzer

Arnaud Blondel

Stefan Boresch

John Brady

Bernard Brooks

Charles L. Brooks, III

Thomas H. Brown

Robert E. Bruccoleri

Paul W. Brumer

Axel T. Brünger

Rafael P. Brüschweiler

Matthias Buck

Amedeo Caflisch

William J. Campion

William Carlson

David A. Case

Leo Caves

Thomas C. Caves

Marco Cecchini

John-Marc Chandonia

Ta-Yuan Chang

Xavier Chapuisat

Sergei Chekmarev

Rob D. Coalson

François Colonna-Cesari

Simone Conti

Michael R. Cook

Qiang Cui

Tara Prasad Das

Annick Dejaegere

Philippe Derreumaux

Aaron Dinner

Uri Dinur

Manvendra K. Dubey

Roland L. Dunbrack, Jr.

Chizuko Dutta

Nader Dutta

Claus Ehrhardt

Ron Elber

Marcus Elstner

Byung Chan Eu

Jeffrey Evanseck

Erik Evensen

Jeffrey Evenson

Thomas C. Farrar

Martin J. Field
Stefan Fischer
David L. Freeman
Thomas Frimurer
Kevin Gaffney
Jiali Gao
Yi Qin Gao
Bruce Gelin
R. Benny Gerber
Paula M. Getzin
Debra A. Giammona
Martin Godfrey
Andrei Golosov
David M. Grant
Daniel Grell
Peter Grootenhuis
Hong Guo
Ogan Gurel
Robert Harris
Karen Haydock
Russell J. Hemley
Jeffrey C. Hoch
Milan Hodoscek
Gary G. Hoffman
L. Howard Holley
Barry Honig
Victor Hruby
Rod E. Hubbard
Robert P. Hurst
Vincent B.-H. Huynh
Toshiko Ichiye
K. K. Irikura
Alfonso Jaramillo
Tom Jordan
Diane Joseph-McCarthy

Sun-Hee Jung
C. William Kern
William Kirchhoff
Burton S. Kleinman
Gearld W. Koeppl
H. Jerrold Kolker
Yifei Kong
Lewis M. Koppel
J. Kottalam
Felix Koziol
Christoph Kratky
Sergei Krivov
Olga Kuchment
Krzysztof Kuczera
John Kuriyan
Joseph N. Kushick
Peter W. Langhoff
Antonio C. Lasaga
Frankie T. K. Lau
Themis Lazaridis
Fabrice LeClerc
Angel Wai-mun Lee
Irwin Lee
Sangyoub Lee
Ming Lei
Ronald M. Levy
Xiaoling Liang
Carmay Lim
Xabier Lopez
Guobin Luo
Paul D. Lyne
Jianpeng Ma
Alexander D. MacKerell, Jr.
Christoph Maerker
Paul Maragkakis

Marc Martí-Renom

Jean-Louis Martin

Carla Mattos

J. Andrew McCammon

H. Keith McDowell

Jorge A. Medrano

Morten Meeg

Marcus Meuwly

Olivier Michielin

Stephen Michnick

Fredrick L. Minn

Andrew Miranker

Keiji Morokuma

A. Mukherji

Adrian Mulholland

David Munch

Petra Munih

Robert Nagle

Setsuko Nakagawa

Kwango Nam

Eyal Neria

John-Thomas C. Ngo

Dzung Nguyen

Lennart Nilsson

Iwao Ohmine

Barry Olafson

Kenneth W. Olsen

Neil Ostlund

Victor Ovchinnikov

Emanuele Paci

Yuh-Kang Pan

C.S. Pangali

Richard W. Pastor

Lee Pedersen

David Perahia

Robert Petrella

B. Montgomery Pettitt

Ulrich Pezzeca

Richard N. Porter

Jay M. Portnow

Carol B. Post

Lawrence R. Pratt

Martine Prévost

Blaise Prod'hom

Jingzhi Pu

Dagnija Lazdins Purins

Lionel M. Raff

Mario Raimondi

Francesco Rao

Gene P. Reck

Swarna Yeturu Reddy

Walter E. Reiher, III

Nathalie Reuter

Bruno Robert

Peter J. Rossky

Benoît Roux

Andrej Sali

Daniel Saltzberg

Michael Schaefer

Michael Schlenkrich

David M. Schrader

John C. Schug

Klaus Schulten

Eugene Shakhnovich

Moshe Shapiro

Ramesh D. Sharma

Isaiah Shavitt

Henry H.-L. Shih

Bernard Shizgal

David M. Silver

Manuel Simoes
Balvinder Singh
Jeremy Smith
Sung-Sau So
Michael Sommer
Ojars J. Sovers
Martin Spichty
David J. States
Richard M. Stevens
Roland Stote
John Straub
Collin Stultz
Neena Summers
Henry Suzukawa
S. Swaminathan
Attila L. Szabo
Antoine Taly
Kwong-Tin Tang
Bruce Tidor
Hideaki Umeyama
Arjan van der Vaart
Wilfred van Gunsteren
Herman van Vlijmen

Michele Vendruscuolo
Dennis Vitkup
Mark Wagman
Shunzhou Wan
Iris Shih-Yung Wang
Ariel Warshel
Masakatsu Watanabe
Kimberly Watson
David Weaver
Paul Weiner
Michael A. Weiss
Joanna Wiórkiewicz-K.
George Wolken
Youngdo Won
Yudong Wu
Robert E. Wyatt
Wei Yang
Robert Yelle
Darrin York
Hsiang-ai Yu
Guishan Zheng
Yaoqi Zhou
Vincent Zoete

Martin Karplus: LIFE IN COLOR—From the 1940s to 2019

"The Visible Image Represents an Invisible Truth"
(from a tenth-century manuscript)

When I was completing my PhD at Cal Tech in 1953, my parents gave me a Leica IIIC. I brought the Leica with me to Oxford, England, where I had a postdoctoral fellowship at the Mathematical Institute. Since I was only 23 years old and had studied continuously all the way through graduate school, I was eager for the sojourn in Europe to provide experiences beyond science. A National Science Foundation fellowship gave me a generous (at the time) salary of $3000 per year, which was sufficient to do considerable traveling. Outside of the three six-week university terms during which I was in residence in Oxford, I made numerous trips throughout Europe. Meeting people and being exposed to their cultures, art, architecture, and cuisines was an incredible experience, which had a lasting effect on my life.

Throughout these travels, I was intent on recording what I saw and took many photographs. They preserve my vision of a world, much of which no longer exists. Economic development, universal communication, and war have taken a heavy toll. Many of the towns and villages have been destroyed or replaced, everyday costumes of the time are worn only at events for tourists, and much of the social fabric of the communities has been destroyed. Many of the people I photographed belonged to the last generation to live in a way that had lasted for centuries. The areas I visited in Europe and America had their own traditions, many of which have now disappeared as the world has been homogenized.

To obtain these images, particularly those of people, I used a Hector long-focus lens. Its reflex viewer made possible recording an image while facing away from the subject. This permitted me to take close-ups of individuals and crowds without their being cognizant of what I was doing. Much later I learned that Paul Strand and Walker Evans used the same stratagem in the early 1920 to obtain some of their famous black-and-white images.

During the academic year 1999–2000, I found myself again living in Oxford, as Eastman Professor. While there, I was introduced to an excellent photographic craftsman, Paul Sims (Colourbox Techunique), who scanned some of the slides so that digital exhibition prints could be made. The first exhibition of the prints was held at The Cloisters (National Institutes of Health) in Bethesda, Maryland (April 2005). This was followed by solo exhibitions at the Carpenter Center for Visual Arts at Harvard University (November 2005), at the Panopticon Gallery in Boston (August, September 2006), and at the Wolfson College Gallery in Oxford, England (November, December 2006). A solo exhibition took place at Stimultania in Strasbourg, France (May, June, 2008) and at the Cambridge Multicultural Center in Cambridge, Massachusetts (November, December, 2008). Images from the collection were projected at the Transphotographique Festival in Lille, France (Spring, Summer 2009) and in a group show at the Real Colegio Complutense in Cambridge, Massachusetts (April, May, 2011). A solo exhibition was held at the Stoneham Theatre Gallery of the Griffin Museum of Photography in Stoneham, Massachusetts (January–March, 2012), at The Gallery of the Hallmark School of Photography in Turner Falls, Massachusetts (April–June, 2012), and at the Multicultural Art Center in Cambridge, Massachusetts (August–December, 2012). A large solo exhibition took place at the Bibliotheque Nationale in Paris, France (May–August, 2013), before my Nobel Prize in Chemistry was awarded later in the year. Other exhibitions have taken place at the Einstein Gallery in Berlin, Germany (October–December, 2014), at the Austrian Cultural Forum, New York City, New York (September–December, 2014), and at the Austrian Cultural Forum, Washington, DC (January–March, 2015). Since then the Bibliothèque Nationale de France (BnF) exhibition has been shown in Vienna, Austria (May–July, 2015), Milan, Italy (April–May, 2016), San Sebastian, Spain (September–November, 2016), and Venice, Italy (October–November, 2018). An exhibition at the Center for European Studies at Harvard focusing on the former Yugoslavia is taking place from October, 2019 through January, 2020. There is a permanent exhibition in the Atrium of Institut de Science et d'Ingenierie Supramoleculaires (ISIS) in Strasbourg, where I worked for many years.

I now use a digital camera, a Canon EOS 70D. It is an excellent camera but does not have the lens of a Leica nor the resolution and color quality of Kodachrome.

I have concentrated on projects in China (2008, 2015) and India (2009), where corresponding changes are now taking place and impinging on a cultural heritage that has existed for generations. During 2015, I worked on projects in Tibet, Cuba, and Morocco.

Nobel Lecture

————— ✦ —————

Angewandte
    Nobel Lectures

M. Karplus

DOI: 10.1002/anie.201403924

Molecular Modeling

Development of Multiscale Models for Complex Chemical Systems: From H + H₂ to Biomolecules (Nobel Lecture)**

Martin Karplus

computational chemistry · laws of motion ·
molecular mechanics · multiscale models ·
potential surfaces

> " *Do not go where the path may lead, go instead where there is no path and leave a trail.*
> *Ralph Waldo Emerson* "

Paraphrasing Ralph Waldo Emerson, a 19th century New England philosopher and essayist, I shall try to show in this lecture how I have gone where there was no path and left a trail. It leads from trajectory studies of the reactions of small molecules to molecular dynamics simulations of macromolecules of biological interest.

In developing computational methods to study complex chemical systems, the essential element has been to introduce classical concepts wherever possible, to replace the much more time-consuming quantum mechanical calculations. In 1929[1] Paul Dirac (Nobel Prize in Physics, 1933) wrote (Figure 1) the now familiar statement: "*The underlying*

Quantum Mechanics of Many-Electron Systems

"The underlying physical laws necessary for the mathematical theory of a large part of physics and the whole of chemistry are thus completely known, and the difficulty is only that the exact application of these laws leads to equations that are much too complicated to be soluble. **It therefore becomes desirable that approximate practical methods of applying quantum mechanics should be developed, which can lead to explanation of the main features of complex atomic systems without too much computation.**"

Figure 2. Continuation of quote from P. A. M. Dirac in 1929 (Ref. [1]).

Quantum Mechanics of Many-Electron Systems

"The underlying physical laws necessary for the mathematical theory of a large part of physics and the whole of chemistry are thus completely known, and the difficulty is only that the exact application of these laws leads to equations that are much too complicated to be soluble."

Figure 1. Quote from P. A. M. Dirac in 1929 (Ref. [1]).

physical laws necessary for the mathematical theory of a large part of physics and the whole of chemistry are thus completely known, and the difficulty is only that the exact application of these laws leads to equations that are much too complicated to be soluble."

However, the paragraph goes on to a less familiar part (Figure 2): "*It therefore becomes desirable that approximate practical methods of applying quantum mechanics should be developed, which can lead to an explanation of the main features of complex atomic systems without too much computation.*" This statement could be regarded as the *leitmotif* of this year's Nobel Prize in Chemistry, but actually Dirac's

paper refers not to introducing classical mechanics, but rather to simplifying the quantum mechanical approaches.

To develop methods to study complex chemical systems, including biomolecules, we have to consider (Figure 3) the two elements that govern their behavior: 1) The potential surface on which the atoms move; and 2) the laws of motion that determine the dynamics of the atoms on the potential surfaces.

The Nobel Prize focused on the development of models for the potential surface. When I visited the Lifson group in 1969, there was considerable excitement about developing empirical potential energy functions primarily for small molecules. The important "new" idea was to use a functional

[*] Prof. M. Karplus
    Department of Chemistry & Chemical Biology, Harvard University
    Cambridge, MA 02138 (USA)
    and
    Laboratoire de Chimie Biophysique, ISIS, Université de Strasbourg
    67000 Strasbourg (France)

[**] Copyright© The Nobel Foundation 2013. We thank the Nobel Foundation, Stockholm, for permission to print this lecture.

Supporting information for this article is available on the WWW under http://dx.doi.org/10.1002/anie.201403924.

© 2014 Wiley-VCH Verlag GmbH & Co. KGaA, Weinheim

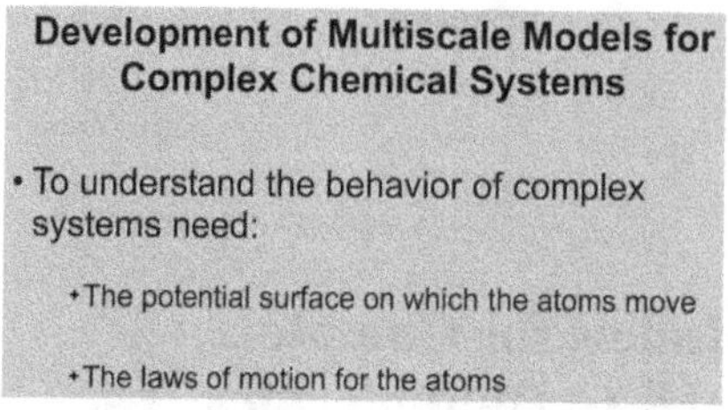

Nobel Lecture

Development of Multiscale Models for Complex Chemical Systems

- To understand the behavior of complex systems need:
  - The potential surface on which the atoms move
  - The laws of motion for the atoms

Figure 3. Essential elements for calculating the behavior of complex chemical systems.

form that could serve not only for calculating vibrational frequencies, as did the expansion of the potential about a known or assumed energy minimum, but also for determining the molecular structure at the minimum. This approach gave rise to molecular mechanics or force fields, as they are now called, in which the energy is expanded in terms of empirical functions that are easy to calculate; the groups of Allinger,[2] Scheraga,[3] and Lifson[4] all made important contributions to the development. The possibility of using such energy functions for larger systems, such as proteins, struck me as very exciting, though I did not work on this for a while.

Since Michael Levitt and Arieh Warshel of the Lifson group are here, I will leave further discussion of potential surfaces to them (Figure 4). In what follows I will focus on the classical treatment of the atomic motions, whether in small molecules or large (Figure 5). Although the laws governing the motions of atoms are quantum mechanical, the key realization that made possible the simulation of the dynamics of complex systems, including biomolecules, was that a classical mechanical description of the atomic motions is adequate in most cases.

From my own perspective, this realization was derived from calculations that my group did in the 1960s, when we studied a very simple reaction, the symmetric exchange reaction, $H + H_2 \rightarrow H_2 + H$. As shown in Figure 6 (upper part), this involves the atom H_C colliding with the molecule $H_A–H_B$ with the result that a new molecule $H_B–H_C$ is formed and the atom A escapes. To determine the trajectories describing the reaction, it is necessary (Figure 3) to know the potential surface governing the interactions between the three atoms.

Martin Karplus was born 1930 in Vienna, Austria, and received his Ph.D. 1953 from the California Institute of Technology. He is currently Professeur Conventionné at the Université de Strasbourg, France, and Theodore William Richards Professor of Chemistry, Emeritus, at Harvard University in Cambridge, Massachusetts, USA. Photo: P. Badge.

The Nobel Prize focused on the development of multiscale models for the potential surface.

- The most important approaches for representing the potential surface of complex systems which do not use quantum mechanics (the so-called force fields) were developed in the Allinger, Lifson and Scheraga groups.

- Different representations for the elementary particles were introduced: atoms, residues, and secondary structures, for example.

- To study chemical reactions, the classical force fields were extended to treat part of the system by quantum mechanics, the so-called QM/MM method.

- Since Michael Levitt and Arieh Warshel of the Lifson group are here, I will leave the discussion of that aspect to them.

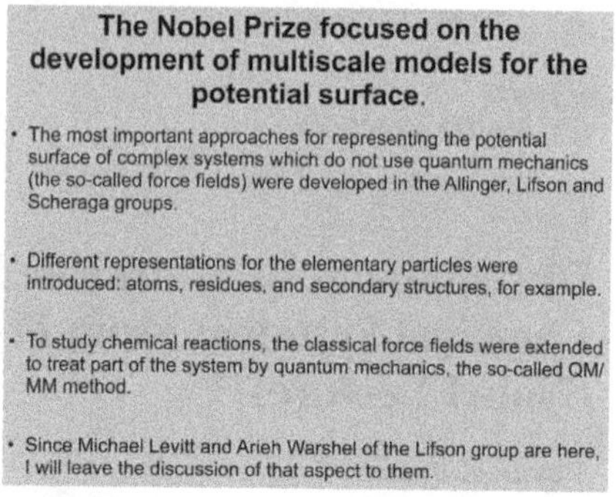

Figure 4. Aspects of potential surface for complex chemical systems.

The laws of motion for the atoms

- Although the laws governing the motions of atoms are quantum mechanical, the essential realization that made possible the treatment of the dynamics of complex systems was that a classical mechanical description of the atomic motions is adequate in most cases

- This realization was derived from simulations of the dynamics of the $H+H_2$ exchange reaction

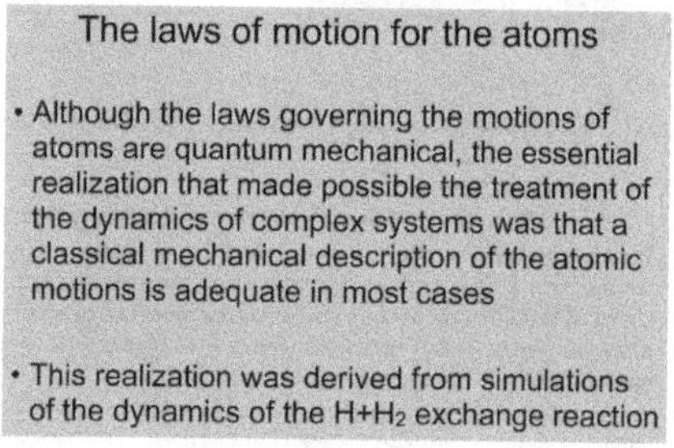

Figure 5. Laws of Motion: Quantum vs Classical

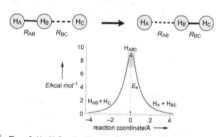

Figure 6. $H + H_2$ Reaction. Upper: colinear reactive collision; Lower: PK potential surface for a colinear reaction (see Ref. [5]).

What Richard Porter and I used was a semi-empirical valence-bond surface.[5] This is not surprising since I had been a student of Linus Pauling (Nobel Prize in Chemistry, 1954; Nobel Prize for Peace, 1962), who believed that valence bond theory was the best approach for understanding chemical bonding. When compared with high-level quantum mechanical calculations,[6] the Porter–Karplus (PK) surface, as it has come to be called, has turned out to be surprisingly accurate, in spite of the simplicity of the approach. The PK

M. Karplus

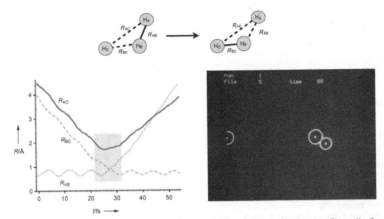

Figure 7. H + H₂ Reactive Collision. Upper: non-colinear reactive collision; Lower-left: atom distances during reactive collision with yellow box indicating the strong interaction region; Lower-right: snapshot of a reactive collision (from Film 1) (see Refs. [8, 36]).

surface has been used by several groups in testing calculational methods for studying the H + H₂ reaction, as described below.[7]

The energy as a function of the reaction coordinate for a colinear collision, which corresponds to the lowest energy reaction path, is shown in the lower part of Figure 6. The essential feature of the surface is that there is a high activation barrier for the reaction. Although Figure 6 shows the colinear surface, the actual trajectories describing the reaction were determined by solving Newton's equation of motion in the full three-dimensional space.[8]

Since there are only three atoms, their relative positions can be described in terms of the three distances between the three pairs of atoms. On the lower left of Figure 7 are shown the distances between the atoms as a function of time in femtoseconds, which is the appropriate timescale for the collision. In this figure, which represents a reactive collision, the distances R_{AC} and R_{BC} decrease as atom H_C collides with molecule H_A–H_B, which is vibrating before the reaction takes place; after the reaction, the newly formed molecule, H_B–H_C, vibrates and atom H_A escapes. The yellow box in the Figure indicates the time during which strong interactions between

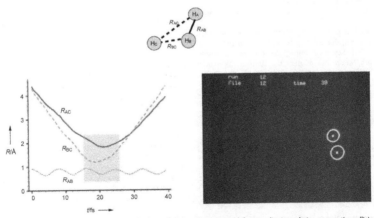

Figure 8. H + H₂ Nonreactive Collision. Upper: non-colinear non-reactive collision; Lower-left: atom distances during nonreactive collision with yellow box indicating the strong interaction region; Lower-right: snapshot of a nonreactive collision (from Film 1) (see Refs. [8, 36]).

the atoms are present; it corresponds to about 10 femtoseconds.

Figure 8 (lower left) shows a nonreactive collision in the same way as the reactive collision is shown in Figure 7. Again, the interaction time (yellow box) is on the femtosecond timescale. In this case, the internuclear distance R_A-R_B continues as a molecule vibration and the colliding atom H_C escapes.

Soon after the calculations were done, Lee Pedersen and Keiji Morokuma, postdoctoral fellows in my group, discovered that there was a graphics laboratory at Harvard and obtained permission to make a film, which shows a series of reactive and non-reactive collisions. A snapshot from the film segments showing a reactive and a nonreactive trajectory are on the lower right of Figures 7 and 8, respectively. A brief description of each of the films is given in the Appendix. The films are available via links given in the Supporting Information.

Even though an individual reaction takes place on the femtosecond timescale, the macroscopic rate is much slower. This difference in timescales arises from the fact that the reaction rate is determined by averaging over a large number of trajectories with an energy distribution corresponding to the Boltzmann Law. Even at a 1000 K, a temperature high enough for the reaction to be easily measured,[9] most of the collisions do not have enough energy to get over the barrier. Consequently, although an individual event is very fast, the overall rate is many orders of magnitude slower.

The classical trajectory calculations of the $H + H_2$ reaction were in approximate agreement with the available experimental data.[9,10] However, it seemed to me important to ascertain that the details of the classical results were correct. For this purpose, it was necessary to have a full quantum mechanical calculation for the $H + H_2$ reaction, which was not available at the time. A significant theoretical development and much more computer time were required. It was only ten years later that a good friend of mine, Aron Kuppermann,[11] and also Bob Wyatt[12] were able to do such a calculation (Figure 9).

Since we had used the approximate PK potential for the classical mechanical calculation, both groups also used the PK potential; i.e., they were testing not whether the results agreed with Nature but whether the classical calculations

were valid. As stated in the figure, they found that the classical results were as accurate as the quantum mechanical results that they obtained with much more work.

The comparison showed that the reaction of hydrogen atoms, for which you would expect the largest quantum effects, can be described classically in most cases. At low temperatures, significant tunnelling can occur, so that quantum corrections are required.[13] Consequently, for heavier atoms, as well as for hydrogen atoms, classical mechanics should be valid for studying the dynamics at ambient temperatures. Since biomolecules are composed mainly of carbon, nitrogen and oxygen, with hydrogen atoms bonded to them, I concluded that classical mechanical molecular dynamics simulations would be meaningful.

Before focusing on the dynamics of larger molecules, I will discuss some work related to one of the papers mentioned in the "Scientific Background" to the Nobel Prize in Chemistry. I had become interested in the chemistry of vision as an undergraduate at Harvard and did research with Ruth Hubbard and George Wald (Nobel Prize in Physiology in 1967). After I returned to Harvard in 1966 as a Professor, I came across an article by Ruth Hubbard and George Wald in a volume dedicated to Linus Pauling for his 65th birthday.[14] It was entitled, "Pauling and Carotenoid Stereochemistry." In it, Hubbard and Wald reviewed Pauling's contribution to the understanding of polyenes with emphasis on the visual chromophore, retinal. The article contained a paragraph, which I reproduce here because it describes an element of Pauling's approach to science that greatly influenced my research:

"One of the admirable things about Linus Pauling's thinking is that he pursues it always to the level of numbers. As a result, there is usually no doubt of exactly what he means. Sometimes his initial thought is tentative because the data are not yet adequate, and then it may require some later elaboration or revision. But it is frequently he who refines the first formulation."

On looking through the article, it was clear to me that the theory of the electronic absorption spectrum of retinal and its geometric changes on excitation, which play an essential role in vision, had not advanced significantly since my discussions with Hubbard and Wald during my undergraduate days at Harvard. I realized, in part from my work in Oxford as a postdoctoral fellow with Charles Coulson, that polyenes, such as retinal, were ideal systems for study by the available semiempirical approaches; that is, if any biologically interesting system in which quantum effects are important could be treated adequately, retinal was it. Barry Honig, who had received his PhD in theoretical chemistry working with Joshua Jortner, joined my research group at that time. He was the perfect candidate to work on the retinal problem.

Figure 10 shows the important conformations of retinal. The active chromophore is 11-*cis*; i.e., the C_{11}–C_{12} double bond is in a *cis* configuration (see Figure 10b). When retinal is photoisomerized, the initial step of vision, it is transformed to 11-*trans*; i.e., the C_{11}–C_{12} double bond is isomerized from *cis* (Figure 10b and c) to *trans* (Figure 10a). In the 11-*cis* state, it is possible to have the two isomers: 11-*cis*,12-s-*cis* (i.e., the C_{12}–C_{13} single bond is *cis*, Figure 10b) and 11-*cis*, 12-s-*trans*

Accurate Quantum Dynamics Treatment of H+H₂ Reaction

- The full QM results "agree with quasiclassical trajectory results of KPS within accuracy of the quantum calculation."

- If Newtonian classical mechanics works for the lightest atom, it should be valid for C, N, O, of which most biomolecules are composed.

Figure 9. Importance of an accurate quantum treatment for validating the classical treatment (see Ref. [8,11]).

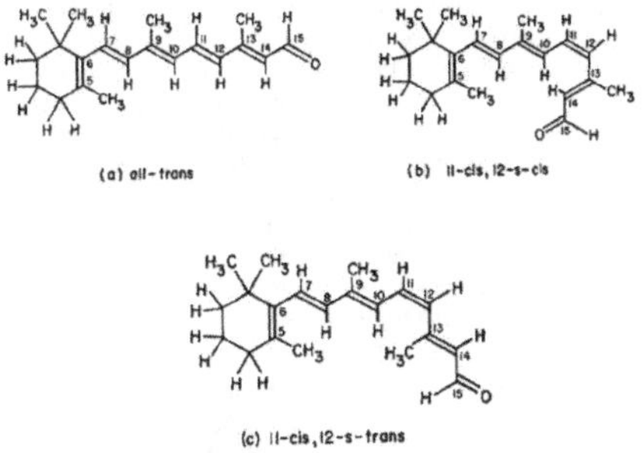

M. Karplus

(a) all-trans

(b) 11-cis, 12-s-cis

(c) 11-cis, 12-s-trans

Figure 10. Retinal conformers. a) all-*trans*: the stable conformer after absorption of light and photoisomerization; b) 11-*cis*,12-s-*cis*: one possible photoactive conformer; c) 11-*cis*,12-s-*trans*: the other possible photoactive conformer (from Ref. [15]).

(Figure 10c). From looking at the two conformers, one would guess that the 12-s-*cis* conformer would be significantly lower in energy, because the H_{10} and H_{14} hydrogens, which appear close enough to repel each other (see Figure 10b) than H_{10} and $(CH_3)_{13}$ (see Figure 10c), which would be expected to have a greater repulsion.

However, when Barry Honig and I calculated the energies in the first paper[15] that used a quantum mechanical model for the π-electrons and a pairwise nonbonded van der Waals interaction energy for the σ-bond framework, we found that the two conformers are very close in energy because the larger expected repulsion in 12-s-*trans* can be reduced significantly by twisting around the single bonds; the difference is only about 1.5 kcal mol^{-1}, with 12-s-*cis* lower. Since these and other results in the paper had significant implications for the visual cycle, we submitted the paper describing them to *Nature*. It received excellent reviews, but came back with a rejection letter stating that because there was no experimental evidence to support our results, it was not certain that the conclusions were correct. This was my first experience with *Nature* and with the difficulty of publishing theoretical results related to biology, particularly in "high impact" journals. The problem is almost as prevalent today as it was then; i.e., if theory agrees with experiment it is not interesting because the result is already known, whereas if one is making a prediction, then it is not publishable because there is no evidence that the prediction is correct. I was sufficiently upset by the editorial decision that I phoned John Maddox, the Editor of *Nature*, and explained the situation to him. Apparently, I was successful, as the paper was finally accepted. Fortunately for Maddox and for us, about six months later, an X-ray structure by Jerome Karle (Nobel Prize in Chemistry 1985) and co-workers[16] was published which confirmed our results. In a review of studies of the visual chromophore,[17] we noted that *"Theoretical chemists tend to use the word 'prediction' rather loosely to refer to any calculation that agrees with experiment, even when the latter was done before the former; the 12 s-cis geometry was a prediction in the true meaning of the word."*

While Arieh Warshel was a postdoctoral fellow in my group, we extended the mixed quantum/classical mechanical method introduced in reference [15] to calculations of the spectrum and vibrations of retinal[18] and similar molecules. This was followed by the use of classical trajectories of the type employed for $H + H_2$ with a simple surface crossing model for the treatment of the photoisomerization process.[19] Figure 11 (bottom left) illustrates the case that was studied. It

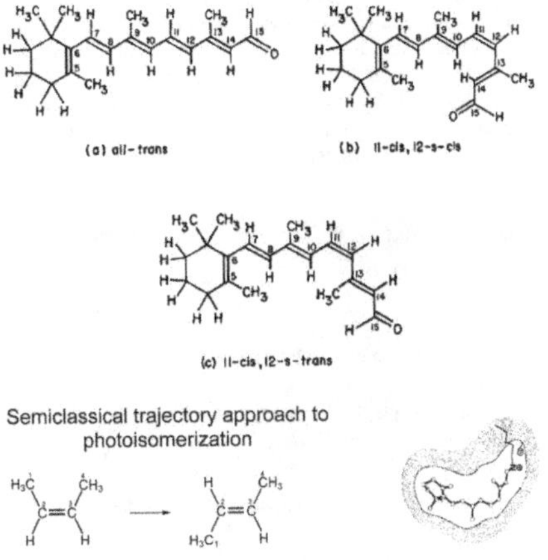

(a) all-trans

(b) 11-cis, 12-s-cis

(c) 11-cis, 12-s-trans

Semiclassical trajectory approach to photoisomerization

Figure 11. Retinal photoisomerization dynamics. Bottom-left: transformation from *cis*- to *trans*-2-butene; bottom-right: suggested constraints on retinal in protein rhodopsin (adapted from Refs. [19,20]).

was the photoisomerization of 2-butene from the cis configuration with the two methyl groups on the same side of the double bond to the trans configuration with the two methyl groups on opposite sides of the double bond.

From looking at Figure 11 (top), it is clear that the photoisomerization of retinal from 11-*cis* to all-*trans*, involves a large displacement of the two ends of the molecule relative to each other for both 12-s-*cis* and 12-s-*trans*. Shortly after Warshel left my group, he published a paper[20] based on the idea that when bound to the protein rhodopsin in the rods of the eye, the ends of the molecule would be restricted from moving significantly during the isomerization. As indicated in Figure 11 (lower right), the model used fixed end groups. To allow the retinal to isomerize without movement of the end groups, he proposed the so-called "bicycle pedal" model. Of course, the rhodopsin was not included in the calculation (i.e., no protein was present) since its structure was not known at the time. Recent studies[21] have shown that the actual isomerization is more complicated than proposed by Warshel and that relaxation of rhodopsin plays a significant role.

In the same year (1976), J. Andrew (Andy) McCammon, Bruce Gelin, and I did the first calculation applying the classical trajectory methodology to a protein, the bovine pancreatic trypsin inhibitor (BPTI). We chose this protein

Angewandte
International Edition Chemie

because it was small (only 58 residues and only 458 (pseudo) atoms in the extended atom model) and because it was one of the few proteins for which a high resolution crystal structure was available.[22] In the mid-1970s, it was difficult to obtain the computer time required to do such a simulation in the United States; the NSF centers did not yet exist. However, CECAM (Centre Européen de Calcul Atomic et Moléculaire) in Orsay, France, directed by Carl Moser, a person with an unusual vision for the future of computations in science, had access to a large computer for scientific research. In the summer of 1976, a two-month workshop was organized at CECAM by Herman Berendsen. Realizing that the workshop was a great opportunity, perhaps the only opportunity, to do the required calculations, Andy McCammon and Bruce Gelin worked very hard to prepare and test a program to do the molecular dynamics simulation of BPTI (Figure 12). Because of their

---

Bovine Pancreatic Trypsin Inhibitor (9.2 ps)

- Classical mechanical potential function based on the work of Scheraga and Lifson groups

- Classical mechanical dynamics based on generalization of the H+H_2 methodology to a large number of atoms

Figure 12. Methodology of BPTI simulation (see text and Ref. [23]).

---

intense preparatory work, Andy was able to start running the molecular dynamics simulation as soon as he arrived. It was essentially completed at the workshop and published in 1977.[23] It is worth mentioning that during this workshop, stimulated by the description of the BPTI simulation, a number of groups began to use molecular dynamics for studying biomolecules. They include W. F. van Gunsteren and H. J. C. Berendsen, J. Hermans and A. Rahman, and M. Levitt (see CECAM Workshop Report on "Models of Protein Dynamics," Orsay, May 24-July 17, 1976.

We used a potential function developed by Bruce Gelin[24] that was a combination of the Scheraga and Lifson group potential functions. The molecular dynamics simulation of BPTI was an extension of what we had done for H + H_2 from a system of 3 atoms to one of 458 (pseudo) atoms. As mentioned earlier, it was a very natural generalization since the classical equations of motions should be applicable, regardless of the number of atoms. It is also important to remember that the BPTI simulation was not the first simulation for a many-particle system with a realistic potential function for the interactions. In particular, Aneesur Rahman, a pioneer in the simulation field who unfortunately died young, had studied liquid argon in 1964[25] and liquid water, in a collaboration with Frank Stillinger in 1974.[26] They seem not to have been concerned with the validity of classical mechanics for these systems; perhaps I was overly cautious. The 9.2 ps simulation of BPTI[23] gave results concerning the fluid-like internal motions of proteins that contrast

sharply with the rigid view inferred from the X-ray structures. The extent of the protein mobility was, in fact, a great surprise to many crystallographers[27] and is one early example of the conceptual insights concerning molecular properties that have been derived from molecular dynamics simulations.

Obviously, the best way to illustrate the motions would have been a film of the trajectory. However, the computer graphics facilities available to us were not advanced enough to treat a 458 (pseudo)-atom system in a finite time. Instead, Bruce Gelin made two drawings of the structure of BPTI (Figure 13), one at the beginning of the simulation (left) and

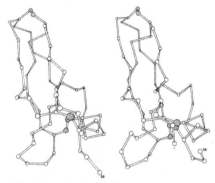

Figure 13. BPTI simulation. Left: Initial structure; right: structure after 3.2 ps. The C_α carbons are indicated by circles, the sulfurs in disulfide bonds by stippled circles, the C_α carbons are connected by rods (from Ref. [23]).

the other (right) after 3.2 picoseconds. If you look carefully at the figure, you can see that although the two structures are very similar, every residue has moved by a small amount. Given that computer graphics can now make the desired film of the trajectory very easily, Victor Ovchinnikov, a postdoctoral fellow in my group, produced a film for the Nobel Lecture using the corresponding representation (see Figure 14 and Film 2).

In an oral history that Andy McCammon recorded in 1995,[28] he made the prescient statement (Figure 15): "*There was a sense, even at the time, of something truly historic going on, of getting these first glimpses of how an enzyme molecule, for example, might undergo internal motions that allow it to function as a biological catalyst.*"

Today, when thousands of molecular dynamics simulations of biomolecules are being done by hundreds of scientists, it is clear that what we felt at that time was indeed the beginning of a new era in the understanding of biological systems. As computers became faster, one could improve the results, not only by refining the potentials, but also by doing longer simulations of more realistic model systems. At the same CECAM workshop where the first BPTI simulation was done, Peter Rossky and I,[29,30] in a collaboration with Aneesur Rahman, did a simulation of the alanine-

M. Karplus

Figure 14. BPTI simulation. Image for Film 2. Same as Figure 13, except that the disulfide bonds are indicated with yellow circles and the light/dark C_α connectors represent the result of light shining on the image. (Drawing made by Victor Ovchinnikov with VMD.)

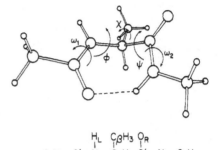

Figure 16. Drawing of alanine dipeptide for the solution simulation. Top: Conformation used in simulation; bottom: chemical formula (Ref. [29]).

> There was a sense, even at the time, of something truly historic going on, of getting these first glimpses of how an enzyme molecule, for example, might undergo internal motions that allow it to function as a biological catalyst.
>
> J. A. McCammon, Oral History (1995)

Figure 15. Based on an interview of J. A. McCammon in 1995, after he received the 1995 Cray Research Leadership Award for Breakthrough Science from the Computer World Foundation (see Ref. [28]).

Simulations of Proteins in Solution

- Simulated BPTI for 210ps in a box of 2,607 water molecules (Levitt & Sharon, '88)

- One millisecond simulation of BPTI in water (Shaw *et al.* 2010)

- So far, no simulation of BPTI folding exists, though smaller protein folding simulations with all-atom models in explicit solvent have been performed (Shaw *et al.* 2011)

Figure 17. Summary of BPTI solution simulations (see text).

dipeptide (Figure 16) in a box of water molecules and showed that the water around the hydrophobic methyl groups behaved differently from the water interacting with the polar C=O and N−H groups.

In 1988 Michael Levitt and Ruth Sharon[31] published a simulation of BPTI (see Figure 17) that was more than twenty times longer than the original simulation and very importantly, the simulation was done in a box of water molecules. The Levitt-Sharon simulation confirmed the water behavior observed in the Rossky et al. papers.[29,30] Further, the simulation was qualitatively in agreement with the original BPTI vaccum simulation results, although the motions of the residues were somewhat smaller and because of the water friction, they were also slightly slower. Recent work[32,33] has elaborated our understanding of the role of the water environment in protein dynamics.

In 2010, Shaw and his co-workers[34] (Figure 17) performed a 1 millisecond simulation of BPTI described by a standard force field using a specially designed computer. The paper analyzed the long time dynamics in detail, but for me the most important aspect of the simulation is that they

found that BPTI was stable on the millisecond timescale. I had always wondered, perhaps been "scared" is a better word, whether with the relatively crude potentials we were using the protein would fall apart (denature) if the molecular dynamics simulations were extended to such long times, the timescales that are of interest for many biological processes.

In relation to such considerations, I would like to remind the audience that a very difficult problem in the field of molecular dynamics simulations of biomolecules is to have a way of checking that the results are correct. Experimental data (e.g. NMR measurements) that can be used for validation of the results are important but limited; i.e., they do not provide enough information for a quantitative test. Despite what the Nobel Prize press citation implies ("The computer is just as important as the test tube."), experiments are essential to verify that what we are doing is meaningful. It is often possible to verify that the statistical error is sufficiently small that the simulations can be used to understand the phenomenon being studied,[35] but the systematic error due to the approximations in the potentials is difficult to quantify.

 © 2014 Wiley-VCH Verlag GmbH & Co. KGaA, Weinheim *Angew. Chem. Int. Ed.* **2014**, *53*, 9992–10005

Angewandte
International Edition *Chemie*

In addition to the dynamics of the native proteins like BPTI, how the polypeptide chain folds to the native state is of great interest.[36] No folding simulation of BPTI is available as yet (Figure 17), though such simulations have been performed for smaller proteins.[37] The present status of our knowledge of BPTI folding, which was first studied by Levitt and Warshel with an ultra-simplified model,[38] is summarized in Ref. [39].

An early example of "multiscale" modeling, in the sense emphasized by the Nobel Prize citation, is the diffusion–collision model for protein folding, which was developed in 1976 by David Weaver and me.[40] It used a coarse-grained description of the protein with helices as the elementary particles, and it showed how the search problem for the native state could be solved by a divide-and-conquer approach. Formulated by Cy Levinthal, the so-called Levinthal Paradox points out that to find the native state by a random search of the astronomically large configuration space of a polypeptide chain would take longer than the age of the earth, while proteins fold experimentally on a timescale of microseconds to seconds. In addition to providing a conceptional answer to the question posed by Levinthal, the diffusion–collision model made possible the estimation of folding rates. The model was ahead of its time because data to test it were not available. Only relatively recently have experimental studies demonstrated that the diffusion–collision model describes the folding mechanism of many helical proteins,[41] as well as some others.[42]

In the lecture so far, I have focused on the history of molecular dynamics simulations of proteins and the qualitative insights about protein motions that were obtained from them. An essential conclusion from the early work, as already mentioned, is that fluid-like internal motions occur in proteins at room temperature. Like so many things that occur naturally, Nature is likely to have made use of them by evolutionary developments. The importance of the internal motions is encapsulated in the now very well-known statement (Figure 18): "… *everything that living things do can be understood in terms of the jigglings and wigglings of atoms*".[43] However, I was amazed when I first found that 2000 years earlier, a Roman poet, Titus Lucretius, who is known for only one poem, "De Rerum Natura", made the following state-

> **"The atoms are eternal and always moving. Everything comes into existence simply because of the random movement of atoms, which, given enough time, will form and reform, constantly experimenting with different configurations of matter from which will eventually emerge everything we know..."**

Figure 19. A rendition by Stephen Greenblatt of Titus Lucretius "The Way Things Are: De Rerum Natura" (Vol.I:1023ff), based on the translation of the poem by Martin Ferguson Smith (Hacket Publishing Co., Cambridge, 2001).

ment (Figure 19): "*The atoms are eternal and always moving. Everything comes into existence simply because of the random movement of atoms, which given enough time, will form and reform constantly experimenting with different configurations of matter from which will eventually emerge everything we know...*"

Titus Lucretius based his poem on the detailed atomic theory of matter developed by the Greek philosopher Democritus (about 400 BC). It distinguishes, for example, the bonding between atoms in liquids and solids. The atomic theory of matter apparently was lost for hundreds of years and revived in Europe only in the 1800s by John Dalton.

These quotations raise the question as to how Nature through evolution has developed the structures of proteins so that their "jigglings and wigglings" have a functional role. As Figure 20 indicates, there are two aspects to this. First, evolution determines the protein structure, which in many cases, though not all, is made up of relatively rigid units that are connected by hinges. They allow the units to move with respect to one another. Second, there is a signal, usually the binding of a ligand, that changes the equilibrium between two structures with the rigid units in different positions.

As an example, I will briefly discuss adenylate kinase, an enzyme which has two major conformations (Figure 21). Its

> **"…everything that living things do can be understood in terms of the jigglings and wigglings of atoms."**

Figure 18. Top: Quote from "Feynman Lectures" (see Ref. [43]); bottom: Richard Feynman (Nobel Prize in Physics, 1965) playing bongo drums (from http://www.richard-feynman.net/index.htm).

> **Putting to work the "Jigglings and Wigglings"**
>
> A) Semirigid domains with hinges
>
> B) Binding of ligand to change equilibria amongst conformations

Figure 20. How the "Jigglings and Wigglings" in the Feynman quote are used by Nature (as interpreted in this lecture).

M. Karplus

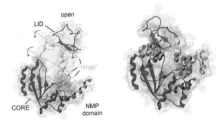

$$2\text{A-P-P} \; \rightleftharpoons \; \text{A-P-P-P} + \text{A-P}$$

Figure 21. Cartoon of adenylate kinase. Left: open structure with no bound substrate showing the hinges; right: closed structure with two bound adenosine diphosphates (A-P-P) (prepared by Victor Ovchinnikov with VMD).

function is to transfer one phosphate group from adenosine diphosphate (A-P-P) to another A-P-P to produce adenosine triphosphate (A-P-P-P) and adenosine monophosphate (A-P). On the left of the Figure is shown the open structure, which permits the substrates to come in and the product to go out, and on the right is shown the closed structure. The closed structure creates a reaction "chamber", which is isolated from the solvent and has the catalytic residues in position for the reaction to take place. Figure 22 (top) shows a series of

a)

b)

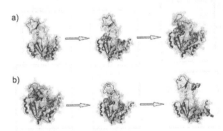

Figure 22. Snapshots from adenylate kinase film (Film 3). a) Closing of enzyme as substrates bind; b) reaction of substrates and opening for product release (prepared by Victor Ovchinnikov with VMD and FFMPEG).

snapshots from a cartoon movie (see Film 3) with the substrate coming in and the enzyme closing; Figure 22 (bottom) shows the reaction taking place and the enzyme opening up to allow the products to escape.

This type of conformational change occurs in many enzymes as an essential part of their mechanism. Moreover, in adenylate kinase and many other enzymes, the chemistry has been optimized such that it is not the rate-limiting step for the overall reaction.[44,45] Jeremy Knowles[46] has called such

enzymes "perfect" since there is no rationale for evolution to further optimize the chemistry when the opening of the enzyme to let the products escape is rate-limiting.

Molecular motors are the prime examples of how the "jigglings and wigglings" are put to work to do something that is essential for life (see Figure 23). My group has studied several different motors, including myosin V,[47,48] F_1 ATPase,[49,50,51] and kinesin.[52,53] I will talk just about one of them, kinesin, because of its relation to this year's Physiology or Medicine Prize, which was awarded for the "discoveries of machinery regulating vesicle traffic, a major transport system in the cell". The work was concerned with genetic analyses of how vesicles open to discharge their cargo at the right time in the right place. Although not all vesicles need to be moved from one place to another, the kinesins, which were discovered in 1982 in the giant squid axon,[54] are very important in the function of many vesicles. The kinesins transport vesicles large distances along the microtubule cytoskeleton of the cell.

Figure 24 shows a set of snapshots from a film (see Film 4) that illustrates how kinesin functions. The two globular "feet" are visible. Actually there are two molecules, each with a globular foot. They are joined together by protein strands, one from each molecule (see also Figure 25), to form a coil-coil at the top of which the vesicle is carried. We know very little about the structure of the vesicles or how they are attached at the top of the coiled-coil. Our research is concerned with understanding the mechanism by which the kinesin dimer walks along the microtubule cytoskeleton. If you look carefully at Film 4, you can see that kinesin walks in the same way as we do: it puts the left foot forward, then the right foot forward, and so on. However, as the film shows the molecules do not walk "normally". The way they walk is like a person who has artificial legs. When you consider the complex muscular and nervous system involved in our walking, how kinesin walks still appears amazing, at least to me.

To understand the walking mechanism, Wonmuk Hwang, Matt Lang and co-workers, and I[52] have been doing molecular dynamics simulations. The snapshots from the film (Figure 24) show that the molecule ATP and its hydrolysis products, ADP and Pi are involved in the stepping mechanism. It is the binding of ATP that trigger the motion by which the back "foot" is "thrown" forward to take a step on the microtubule. To examine the mechanism in more detail, the X-ray structure of a kinesin dimer, shown in Figure 25, was used as the basis for the simulations.[56] Calculations showed that the β-strand, labeled β_{10} in the figure, which serves as the connector, is not sufficiently rigid to be able to perform the so-called "power-stroke", in which the back foot is thrown forward. We noticed that there was another β-strand, labeled β_0, at the N-terminus of the molecule. It is disordered in certain structures, but in others it forms a two-stranded β-sheet with β_{10}. We called β_0 the "cover strand" (CS) and the two-stranded β-sheet, the "cover-neck bundle" (CNB).

Figure 26 shows a pictorial representation of the simulation results. In each of the three diagrams on the left we can see the two feet with a model of the microtubule below. In the

© 2014 Wiley-VCH Verlag GmbH & Co. KGaA, Weinheim

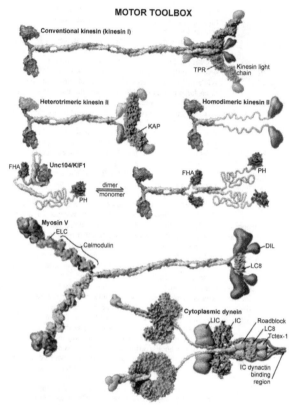

MOTOR TOOLBOX

Figure 23. Cartoon of different types of molecular motors (see R. D. Vale, *Cell* **2003**, *112*, 467–480 for details concerning the image). Many of the cellular elements are motors: they convert energy into motion. Myosin walks on actin, and kinesins walk on microtubules to transport materials in an organized fashion. One motor, F_0F_1-ATP synthase, is special: it does not walk but is resposible for the synthesis of ATP.

ure 27a). Figure 27b shows a cartoon of the experiment. Figure 27c presents one set of results, namely the decrease of the stall force required for the G2 mutant and the almost zero stall force required for DEL, which appears at best to "limp" along the microtubule; more details of the experimental studies that support the CNB model are described separately.[53] Additional simulations are in progress to increase our understanding of how kinesins function. An essential element that is being investigated concerns the role of the interactions between kinesin and the microtubule in the walking mechanism.

Kinesin motors, like other molecular motors, are very important in making life possible.[57] As indicated in Figure 28, mitosis and cell division are inhibited when kinesins do not function due to deleterious mutations. Their importance in cell division makes them a target for cancer chemotherapy. Kinesins are also essential for axonal transport where material has to be delivered over long distances. Some viruses have learned that if they attach themselves to kinesins where the normal cargo would be located, they are transported along the microtubules from one part of the cell to another in a few minutes instead of the ten or so hours that would be required by diffusion in the complex cellular medium.

What does the future hold (Figure 29)? All of us know that real predictions are hard, so I have included relatively conservative ones in the figure. The first, which was mentioned in the introduction, has been a dream of mine since I began to do biomolecular simulations. It is not that simulations can replace all experiments, as the Nobel press announcement seems to imply, but rather that experimentalists would use simulations as a tool like any other (such as X-rays or NMR) in their work to get a better understanding than they could derive from either experiments or simulations alone. That experimentalists are beginning to employ simulations in this way is evidenced by the literature.[58] The respectability for molecular dynamics simulations provided by the Nobel Prize is likely to increase their utilization by the scientific community.

In terms of actual simulations, people are studying more complicated systems. They are beginning to use molecular dynamics simulations for viruses, ribosomes, and even cells so as to gain insights into how they function. If I were thirty years younger I would be simulating the brain. About twenty years

top diagram (A) the forward foot has a disordered cover strand in blue. When ATP binds, the simulations show (middle panel (B)) that the two-stranded cover-neck bundle is formed. It looks very much like a spring and appears to be a high-energy construct. Simulations suggest that, in fact, it acts like a spring with a forward bias that generates the power stroke by propelling the back foot forward (bottom panel (C)) in readiness for the next step.

To test the model based on the simulations, optical trapping experiments in the presence of an external force were performed for a wild-type kinesin and for two mutants.[53] One set of mutations introduced two glycines (G2), which are expected to make the CNB more flexible and the other completely deleted the cover strand (DEL) (Fig-

© 2014 Wiley-VCH Verlag GmbH & Co. KGaA, Weinheim

M. Karplus

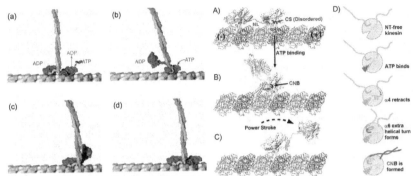

Figure 24. Kinesin walking. Snapshots from Film 4 (created by Graham Johnson for R. D. Vale and R. A. Milligan, 2000; see Ref. [55]). a) View of two globular domains (the "feet") bound to a microtubule; ADP has been released and ATP is binding to the front foot, triggering the powerstroke (see Figure 26 and text); b) release of rear foot; c) partly complete powerstroke; d) completed step.

Figure 26. Schematic representation of the generation of the power-stroke based on the simulations. A) Before ATP binding; B) after ATP binding; C) power stroke; D) diagram highlighting the major molecular events leading to CNB formation and the power stroke (see Ref. [53] and text).

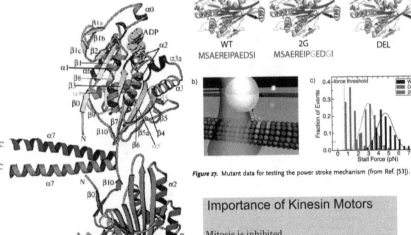

Figure 27. Mutant data for testing the power stroke mechanism (from Ref. [53]).

Figure 25. X-ray structure of rat brain kinesin dimer. The β_{10} strand of each monomer connecting to the coiled-coil and the β_0 strand which is the CS are evident (from Ref. [56]).

Importance of Kinesin Motors

Mitosis is inhibited.

Physiological cargoes are not delivered appropriately (e.g. clogging of axonal transport).

Non-physiological cargoes make use of the transport system (e.g. viruses).

Figure 28. Importance of kinesin motors.

Nobel Lecture

Angewandte
International Edition Chemie

What does the future hold?

• Experimentalists use simulations as a tool like any other

• Applications of simulations to ever more complex systems (viruses, ribosomes, cells, the brain, ...)

Always with cautionary realization that simulations, like experiments, have their limitations and inherent errors.

Figure 29. Future of molecular dynamics simulations.

ago, I spent a couple of years learning what was known about the brain and concluded that not enough data were available to permit me to contribute significantly by making studies on the molecular level. I do not regret the time spent in this way

since I learned much of interest and my research group continued to focus on problems that we could solve. Our knowledge of the brain has increased sufficiently that I would now urge young scientists to work at this exciting frontier, which is beginning to be probed by initiatives in both Europe and America.

However bright the future, I want to caution the audience (as I always do with my students) that simulations have limitations, just as do experiments. In particular, when you appear to have discovered something new and exciting, you should be doubly careful to make certain that there is no mistake in what you have done. Moreover, the example of my exploration of brain research permits me to make an important point. In working at the interface of chemistry and biology with simulation techniques, it is essential to realize that of the many exciting systems that are being studied experimentally, only relatively few pose questions for which molecular dynamics simulations can provide useful insights at their present stage of development.

Figure 30 lists the people to whom this lecture is dedicated. They are the Karplusians: 244 people who have

Karplusian: 1955-2013

Ivana Adamovic	Qiang Cui	L. Howard Holley	Paul D. Lyne	B. Montgomery Pettitt	David J. States
Yuri Alexeev	Tara Prasad Das	Barry Honig	Jianpeng Ma	Ulrich Pezzeca	Richard M. Stevens
David H. Anderson	Annick Dejaegere	Victor Hruby	Alexander D. MacKerell, Jr.	Richard N. Porter	Roland Stote
Ioan Andricioaei	Philippe Derreumaux	Rod E. Hubbard	Christoph Maerker	Jay M. Portnow	John Straub
Yasuhide Arata	Aaron Dinner	Robert P. Hurst	Paul Maragkakis	Carol B. Post	Collin Stultz
Georgios Archontis	Uri Dinur	Vincent B.-H. Huynh	Marc Marti-Renom	Lawrence R. Pratt	Neena Summers
Gabriel G. Balint-Kurti	Roland L. Dunbrack, Jr.	Toshiko Ichiye	Jean-Louis Martin	Martine Prévost	Henry Suzukawa
Christian Bartels	Chizuko Dutta	K. K. Irikura	Carla Mattos	Blaise Prod'hom	S. Swaminathan
Paul Bash	Nader Dutta	Alfonso Jaramillo	J. Andrew McCammon	Jingzhi Pu	Attila L. Szabo
Donald Bashford	Claus Ehrhardt	Tom Jordan	H. Keith McDowell	Dagnija Lazdins Purins	Antoine Taly
Mark Bathe	Ron Elber	Diane Joseph-McCarthy	Jorge A. Medrano	Lionel M. Raff	Kwong-Tin Tang
Oren M. Becker	Marcus Elstner	Sun-Hee Jung	Morten Meeg	Mario Raimondi	Bruce Tidor
Robert Best	Byung Chan Eu	C. William Kern	Marcus Meuwly	Francesco Rao	Hideaki Umeyama
Anton Beyer	Jeffrey Evanseck	William Kirchhoff	Olivier Michielin	Gene P. Reck	Arjan van der Vaart
Robert Birge	Erik Evensen	Burton S. Kleinman	Stephen Michnick	Swarna Yeturu Reddy	Wilfred van Gunsteren
Ryan Bitetti-Putzer	Jeffrey Evenson	Gearld W. Koeppl	Fredrick L. Minn	Walter E. Reiher III	Herman van Vlijmen
Arnaud Blondel	Thomas C. Farrar	H. Jerrold Kolker	Andrew Miranker	Nathalie Reuter	Michele Vendruscuolo
Stefan Boresch	Martin J. Field	Yifei Kong	Keiji Morokuma	Bruno Robert	Dennis Vitkup
John Brady	Stefan Fischer	Lewis M. Koppel	A. Mukherji	Peter J. Rossky	Mark Wagman
Bernard Brooks	David L. Freeman	J. Kottalam	Adrian Mulholland	Benoît Roux	Shunzhou Wan
Charles L. Brooks III	Thomas Frimurer	Felix Koziol	David Munch	Andrej Sali	Iris Shih-Yung Wang
Thomas H. Brown	Kevin Gaffney	Christoph Kratky	Petra Munih	Daniel Saltzberg	Ariel Warshel
Robert E. Bruccoleri	Jiali Gao	Sergei Krivov	Robert Nagle	Michael Schaefer	Masakatsu Watanabe
Paul W. Brumer	Yi Qin Gao	Olga Kuchment	Setsuko Nakagawa	Michael Schlenkrich	Kimberly Watson
Axel T. Brünger	Bruce Gelin	Krzysztof Kuczera	Kwango Nam	David M. Schrader	David Weaver
Rafael P. Brüschweiler	R. Benny Gerber	John Kuriyan	Eyal Neria	John C. Schug	Paul Weiner
Matthias Buck	Paula M. Getzin	Joseph N. Kushick	John-Thomas C. Ngo	Klaus Schulten	Michael A. Weiss
Amedeo Caflisch	Debra A. Giammona	Peter W. Langhoff	Lennart Nilsson	Eugene Shakhnovich	Joanna Wiórkiewicz-K.
William J. Campion	Martin Godfrey	Antonio C. Lasaga	Dzung Nguyen	Moshe Shapiro	George Wolken
William Carlson	Andrei Golosov	Frankie T. K. Lau	Iwao Ohmine	Ramesh D. Sharma	Youngdo Won
David A. Case	David M. Grant	Themis Lazaridis	Barry Olafson	Isaiah Shavitt	Yudong Wu
Leo Caves	Daniel Grell	Fabrice LeClerc	Kenneth W. Olsen	Henry H.-L. Shih	Robert E. Wyatt
Thomas C. Caves	Peter Grootenhuis	Angel Wai-mun Lee	Neil Ostlund	Bernard Shizgal	Wei Yang
Marco Cecchini	Hong Guo	Irwin Lee	Victor Ovchinnikov	David M. Silver	Robert Yelle
John-Marc Chandonia	Ogan Gurel	Sangyoub Lee	Emanuele Paci	Manuel Simoes	Darrin York
Ta-Yuan Chang	Robert Harris	Ming Lei	Yuh-Kang Pan	Balvinder Singh	Hsiang-ai Yu
Xavier Chapuisat	Karen Haydock	Ronald M. Levy	C.S. Pangali	Jeremy Smith	Guishan Zheng
Sergei Chekmarev	Russell J. Hemley	Xiaoling Liang	Richard W. Pastor	Sung-Sau So	Yaoqi Zhou
Rob D. Coalson	Jeffrey C. Hoch	Carmay Lim	Lee Pedersen	Michael Sommer	Vincent Zoete
François Colonna-Cesari	Milan Hodoscek	Xabier Lopez	David Perahia	Ojars J. Sovers	
Michael R. Cook	Gary G. Hoffman	Guobin Luo	Robert Petrella	Martin Spichty	

Figure 30. List of Karplusians (2013). These are collaborators who have worked with me in Illinois, Columbia, Harvard, Paris, and Strasbourg.

worked in my "laboratory" in Illinois, Columbia, Harvard, Paris and Strasbourg. Without them, I would not be here today. Over the last forty years, many of them have contributed to the methodology and applications of molecular dynamics simulations. In writing this, I find it curious that molecular dynamics simulations were not mentioned in the description of the "Scientific Background" of the Nobel Prize. The large community involved in molecular dynamics simulations, which includes all of this year's Nobel Laureates in Chemistry, has transformed the field from an esoteric subject of interest to only a small group of specialists into a central element of modern chemistry and structural biology. Without molecular dynamics simulations and their explosive development, no Nobel Prize would have been awarded in this area.

There is perhaps a parallel here between the fact that molecular dynamics was not mentioned in the Nobel Prize citation and the citation for Einstein's Nobel Prize in Physics (1921). He was awarded the Nobel Prize for the theory of the photoelectric effect and not for his most important work, the general theory of relativity, which had already been verified by experiment and was the origin of his worldwide fame as a scientist. Interestingly, when he gave his Nobel Lecture, it was on relativity, even though he knew that he was supposed to talk about the photoelectric effect. Correspondingly, I traced the history of molecular dynamics simulations and their development in my lecture and did not emphasize the development of potential functions for simulations, the focus of the Chemistry Nobel Prize citation. The complex deliberations of the Physics Committee in reaching its decision concerning Einstein's Nobel Prize are now known because his prize was awarded more than fifty years ago.[59] The public will again have to wait fifty years to find out what motivated the Chemistry Committee in awarding this year's Nobel Prize.

I very much want to mention one other person, my wife Marci, who was willing to live with me, someone "who spent all his time working", in her words. Even more than just living with me, she was brave enough to be my laboratory administrator. Among many aspects of our life, it made possible our working in both the USA and France over many years. Moreover, in preparing to come to Stockholm, the complexity of arranging to be in the right place at the right time would have been overwhelming if she had not been there to take care of what was needed.

Appendix: Background of Films (see Supporting Information)

Film 1. H + H$_2$ Collisions

The film shows two trajectories, the first reactive (Film 1a) and the second nonreactive (Film 1b). In the non-reactive trajectory, it is evident that one of the atoms in the molecule comes out in front of the plane of the reaction and the other goes into the back of the plane. This is done by introducing perspective; i.e., by having an atom grow larger as it comes forward toward you and become smaller as it goes away from you.

In making the film, a question arose as to how to represent the perspective. If the radius of the atomic circles was varied linearly with the distance in front or in back, the perspective was difficult to perceive. So we had to find a better way of showing the perspective.

What I did was to look at the paintings of Canaletto in visits to Venice, and compare the actual distances with how he presented them in his paintings. I found that he seemed to use an approximate exponential law, e^{aR}, where R is the distance out of the plane and α is a coefficient, whose value I do not remember. If I had published this result (There are many things that I did, which were not published.) perhaps there would be a Karplus Law in art theory, as well as the Karplus Equation in nuclear magnetic resonance.

It is also worth remembering the film is of historical interesting for several reasons. Made in 1967, it is the first film to show pictorially the results of an accurate calculation of the motions of the atoms involved in a chemical reaction. The film was made in the laboratory of Professor Sutherland, who was developing the first computer ray-graphics machine. It was a prototype of the devices now manufactured by Evans and Sutherland, which are used, for example, for air traffic control.

Film 2. BPTI Dynamics

The film shows the dynamics of BPTI over about 10 ps, in correspondence with Figure 14. The film was made by Victor Ovchinnikov with FFMPEG based on the images drawn with VMD.

Film 3. Cartoon: Adenylate Kinase Dynamics

Film 3a shows the closing of adenylate kinase by the hinge-bending motions as the two A-P-P substrates bind, and Film 3b shows the reaction to form A-P-P-P and A-P in the closed molecule followed by opening through hinge-bending motions as the products escape. The film was made by Victor Ovchinnikov with FFMPEG using images prepared with VMD.

Film 4. Cartoon: Kinesin Walking on Microtubules

The film shows kinesin taking several steps on the microtubule (see Figure 24 and text). It was made by the group of R. D. Vale and R. A. Milligan.[55]

Received: April 2, 2014
Published online: July 27, 2014

[1] P. A. M. Dirac, *Proc. R. Soc. London Ser. A* **1929**, *123*, 714–733.
[2] N. L. Allinger, M. A. Miller, L. W. Chow, R. A. Ford, J. C. Graham, *J. Am. Chem. Soc.* **1965**, *87*, 3430–3435.
[3] G. Némethy, H. Scheraga, *Biopolymers* **1965**, *4*, 155–184.
[4] S. Lifson, A. Warshel, *J. Chem. Phys.* **1968**, *49*, 5116–5129.

To download the films, please follow the instructions on page 287.

Nobel Lecture

[5] R. N. Porter, M. Karplus, *J. Chem. Phys.* **1964**, *40*, 1105–1115.

[6] P. Siegbahn, B. Liu, *J. Chem. Phys.* **1978**, *68*, 2457.

[7] G. C. Schatz, *Theor. Chem. Acc.* **2000**, *103*, 270–272.

[8] M. Karplus, R. N. Porter, R. D. Sharma, *J. Chem. Phys.* **1966**, *45*, 3871–3873.

[9] A. Farkas, L. Farkas, *Proc. R. Soc. London Ser. A* **1935**, *152*, 124.

[10] S. Datz, E. H. Taylor, *J. Chem. Phys.* **1963**, *39*, 1896.

[11] A. Kuppermann, G. C. Schatz, *J. Chem. Phys.* **1975**, *62*, 2502.

[12] A. B. Elkowitz, R. E. Wyatt, *J. Chem. Phys.* **1975**, *62*, 2504.

[13] M. Garcia-Viloca, J. Gao, M. Karplus, D. G. Truhlar, *Science* **2004**, *303*, 186–195.

[14] "Structural implications of reaction kinetics": R. Hubbard, G. Wald in *Structural Chemistry and Molecular Biology: a Volume dedicated to Linus Pauling by his Students, Colleagues, and Friends* (Ed.: A. Rich, N. Davidson), Freeman, San Francisco, **1968**, pp. 837–847.

[15] B. Honig, M. Karplus, *Nature* **1971**, *229*, 558–560.

[16] R. Gilardi, I. L. Karle, J. Karle, W. Sperling, *Nature* **1971**, *232*, 187–189.

[17] B. Honig, A. Warshel, M. Karplus, *Acc. Chem. Res.* **1975**, *8*, 92–100.

[18] A. Warshel, M. Karplus, *J. Am. Chem. Soc.* **1972**, *94*, 5612–5625.

[19] A. Warshel, M. Karplus, *Chem. Phys. Lett.* **1975**, *32*, 11–17.

[20] A. Warshel, *Nature* **1976**, *260*, 679–683.

[21] V. R. I. Kaila, R. Send, D. Sundholm, *J. Phys. Chem. B* **2012**, *116*, 2249–2258.

[22] J. Deisenhofer, W. Steigemann, *Acta Crystallogr. Sect. B* **1975**, *31*, 238–250.

[23] J. A. McCammon, B. R. Gelin, M. Karplus, *Nature* **1977**, *267*, 585–590.

[24] B. R. Gelin, M. Karplus, *Proc. Natl. Acad. Sci. USA* **1975**, *72*, 2002–2006.

[25] A. Rahman, *Phys. Rev.* **1964**, *136*, A405–11.

[26] F. H. Stillinger, A. Rahman, *J. Chem. Phys.* **1974**, *60*, 1545–1557.

[27] D. C. Phillips in *Biomolecular Stereodynamics, Vol. II* (Ed.: R. H. Sarma), Adenine Press, Guilderland, New York, **1981**, p. 497.

[28] The Computer World Information Technology Foundation, Oral History Collection Archives, **1995**.

[29] P. J. Rossky, M. Karplus, A. Rahman, *Biopolymers* **1979**, *18*, 825–854.

[30] P. J. Rossky, M. Karplus, *J. Am. Chem. Soc.* **1979**, *101*, 1913–1937.

[31] M. Levitt, R. Sharon, *Proc. Natl. Acad. Sci. USA* **1988**, *85*, 7557–7561.

[32] D. Vitkup, D. Ringe, G. A. Petsko, M. Karplus, *Nat. Struct. Biol.* **2000**, *7*, 34–38.

[33] A. L. Tournier, J. C. Xu, J. C. Smith, *Biophys. J.* **2003**, *85*, 1871–1875.

[34] D. E. Shaw, P. Maragakis, K. Lindorff-Larsen et al., *Science* **2010**, *330*, 341–346.

[35] W. Yang, R. Bitetti-Putzer, M. Karplus, *J. Chem. Phys.* **2004**, *120*, 2618–2628.

[36] C. M. Dobson, A. Sali, M. Karplus, *Angew. Chem.* **1998**, *110*, 908–935; *Angew. Chem. Int. Ed.* **1998**, *37*, 868–893.

[37] K. Lindorff-Larsen, S. Piana, R. O. Dror, D. E. Shaw, *Science* **2011**, *334*, 517–520.

[38] M. Levitt, A. Warshel, *Nature* **1975**, *253*, 694.

[39] J. C. Smith, B. Roux, *Structure* **2013**, *21*, 2102–2105.

[40] M. Karplus, D. L. Weaver, *Nature* **1976**, *260*, 404–406.

[41] M. Karplus, D. L. Weaver, *Biopolymers* **1979**, *18*, 1421–1437.

[42] M. Karplus, D. L. Weaver, *Protein Sci.* **1994**, *3*, 650–668.

[43] R. P. Feynman, R. B. Leighton, M. Sands, *The Feynman Lectures in Physics, Vol. I*, Addison-Wesley, Reading, **1963**, chap. 3.

[44] K. A. Henzler-Wildman, V. Thai, M. Lei, M. Ott, M. Wolf-Watz, T. Fenn, E. Pozharski, M. A. Wilson, G. A. Petsko, M. Karplus, C. G. Hübner, D. Kern, *Nature* **2007**, *450*, 838–844.

[45] K. A. Henzler-Wildman, M. Lei, V. Thai, S. J. Kerns, M. Karplus, D. Kern, *Nature* **2007**, *450*, 913–916.

[46] A. L. Wendell, R. J. Raines, J. R. Knowles, *Biochemistry* **1988**, *27*, 1158–1167.

[47] M. Cecchini, A. Houdusse, M. Karplus, *PLoS Comput. Biol.* **2008**, *4*, e1000129.

[48] V. Ovchinnikov, B. L. Trout, M. Karplus, *J. Mol. Biol.* **2010**, *395*, 815–833.

[49] W. Yang, Y. Q. Gao, Q. Cui, J. Ma, M. Karplus, *Proc. Natl. Acad. Sci. USA* **2003**, *100*, 874–879.

[50] Y. Q. Gao, W. Yang, M. Karplus, *Cell* **2005**, *123*, 195–205.

[51] J. Pu, M. Karplus, *Proc. Natl. Acad. Sci. USA* **2008**, *105*, 1192–1197.

[52] W. Hwang, M. J. Lang, M. Karplus, *Structure* **2008**, *16*, 62–71.

[53] A. S. Khalil, D. C. Appleyard, A. K. Labno, A. Georges, M. Karplus, A. M. Belcher, W. Hwang, M. J. Lang, *Proc. Natl. Acad. Sci. USA* **2008**, *105*, 19247–19251.

[54] S. T. Brady, R. J. Lasek, R. D. Allen, *Science* **1982**, *218*, 1129–1131.

[55] R. D. Vale, R. A. Milligan, *Science* **2000**, *288*, 88–95.

[56] F. Kozielski, S. Sack, A. Marx, *Cell* **1997**, *91*, 985–994.

[57] E. Mandelkow, E. M. Mandelkow, *Trends Cell Biol.* **2002**, *12*, 585–591.

[58] M. A. Young, S. Gonfloni, G. Superti-Furga, B. Roux, J. Kuriyan, *Cell* **2001**, *105*, 115–126.

[59] M. Friedman, *The Politics of Excellence*, Henry Holt and Company, New York, **2001**, chap. 7.

Haaretz Article

———————— ❧ ————————

'Two States in One Land':
A Nobel Prize Chemist's Search for Peace

The separation between Israelis and Palestinians is a heavy obstacle to peace: Working together to improve their common homeland would greatly benefit both Jews and Arabs.

by Martin Karplus
June 16, 2014

Policemen separate Israelis and Palestinians at Damascus gate, Jerusalem Day, May 8, 2013.
Credit: AFP

Partly by chance, partly by design, I became individually involved in the Israeli-Palestinian peace process.

When I met Herbert Kelman, Professor of Social Ethics Emeritus at Harvard University, at a conference held in his honor on "The Transformation of Intractable Conflicts", I learned about his "One Country, Two States" proposal. I later rephrased this as "One Land, Two States". Kelman describes his hope for both Israelis and Palestinians as follows: "The acknowledgement that each people is attached to the entire land even though it claims only part of it for its own independent state may well strike a responsive chord in both publics" and make it easier for both "to accept the compromises entailed by an agreement".

Although I strongly support Kelman's proposal, I realize that the present Israeli leadership is not ready to accept it. It seems to be generally acknowledged that an Israeli leader equal to Yitzhak Rabin, who is trusted by both Jews and Arabs, is needed for making peace. Mahmoud Abbas' acknowledging the existence of the Holocaust was a significant step toward building a trusting relationship.

So it was with an intention to relate to both Israeli and Palestinian concerns that I responded to Bar Ilan university president David Hershkowitz when he invited me to receive an honorary degree: "I am honored by the invitation.... However, I do have a concern...My hope is that having been honored by a Nobel Prize, I can do some good while I am in the limelight. For Israel, it is to further the peace process in any way possible. Of course, an individual can only hope to do small things to bring Israelis and Palestinians

Originally published in Haaretz, June 16, 2014.

together. The West-Eastern Divan Orchestra is a case in point. When I gave a lecture [in Israel] some years ago, I expressed the hope that the next lecture I gave in Israel would be attended by students and faculty from Palestinian universities. Clearly the peace process has not advanced in the intervening years, but I do want to keep my word. I wish you to invite not only people from universities in Israel but also from universities in the Palestinian Territories."

I focused on six Palestinian universities with known scientists and both Bar-Ilan's External Relations vice-president and I personally contacted them. Those who replied wrote that they were honored to be invited but that they were unable to accept. Typical is the following: "But, putting all this [the political situation] aside, I am not personally ready to subject myself to the humiliation to which I would be subjected when crossing a checkpoint into Israel itself." Having myself witnessed what can happen to a Palestinian at such a checkpoint when I was visiting the West Bank, I can empathize with these scientists. Clearly, a more humane approach to the security requirements would aid in improving contacts between scientists in the Palestinian territories and Israel.

Separation barrier credit: Martin Karplus

It was the failure of my initial, perhaps naive, effort to bring Israelis and Palestinians together at Bar-Ilan that led me to explore the possibility of presenting a parallel lecture at a Palestinian university. Through the Israeli Palestine Science Organization (IPSO), I had made contact with Hasan Dweik, Vice President of Al Quds University; together with Professor Khalid Kanan, he invited me to present my lecture at the Abu Dis campus, to which scientists from all six Palestinian universities to which invitations had been sent were present as well as about a hundred students, male and female. After the lecture, we discussed the peace process.

I expressed the hope that my visit as a Nobel Prize winner would make the world aware of the importance of science in the Palestinian territories.

Further, since neither the often discussed "two state solution" nor the "one state solution" appears to be viable - the former cannot be implemented because of the settlement problem and the latter would be the end of the Jewish state of Israel (see, for example, Dov Waxman's Haaretz opinion piece earlier this year, ("Time to choose: Liberalism or Zionism?") I proposed that, instead, Kelman's concept of "One Land and Two States" should be considered. In essence, the present Israel and Palestinian Territories together ("One Land") historically are the homeland of both Jews and Arabs and should be recognized as such. Within the historical homeland there would be two sovereign entities ("Two States") that correspond approximately to the present-day Israel and Palestinian Territories, though the exact borders are one of the many problems that would have to be resolved.

Before the leaders of Israel and the Palestinian Territories would trust each other sufficiently to take joint responsibility for the "One Land", much would have to change. The Vatican prayer meeting between the Israeli and Palestinian presidents, brokered by Pope Francis, is one small step in that direction. Hopefully my actions during this visit to the Middle East will also contribute.

Jews and Arabs working together to improve their common homeland would be greatly beneficial to both. The United Nations declared 2013 the Year of Water Cooperation. Given the shortage of water in the region, using it most efficiently by extending Israel's desalination technology and hydroponic agriculture to the Palestinian Territories would be a way to constructively work together.

With "Two States" in the "One Land", the Jewish settlers in what is now the Palestinian territories could continue to live there and would have the choice of becoming Palestinian residents or citizens, an analogy to the Arabs who live in Israel.

After this discussion, I also had the opportunity to visit the Al-Quds campus and see some research laboratories. One was devoted to virology and was very well equipped with the instruments able to identify which virus is involved in a particular infection. This equipment was bought with an outside grant. Although large donations are made to the research at Al Quds for equipment and buildings, the small sums needed to run an apparatus are much harder to come by. Here collaborations with Israeli hospitals to supply the funds needed to make use of the laboratory for testing could help to build trust.

I also learned about other efforts, often with little publicity, that involve cooperation between Israelis and Palestinians. One such joint venture is in the area of high-tech development companies. The startup cost is small and the success depends primarily on the intelligence and originality of the people that contribute. There also exists the Palestinian-Israeli Research Group (PICR) for Israeli-Palestinian medical cooperation to further the treatment and research on infectious diseases.

At this stage, a focus on cooperation in education, technology, medicine, and economics on a personal and small organizational level can help to improve life in the Palestinian Territories and serve to develop the trust required for political progress.

Given the present political situation, I believe that small steps are the most any individual like myself can take to improve the relation between Jews and Arabs.

Let me end with part of an email I received after my lecture at Bar-Ilan.

"Thank you for the words with which you began your lecture, and thank you for making your appearance at Bar Ilan conditional on the ability of Palestinian colleagues to come hear you. Our inability to make peace with our Palestinian neighbors is, I believe, first and foremost a consequence of being separated from them, and any course of action that brings us together is therefore a blessing. I come from a long line of peace activists, and am convinced that the more we two peoples meet each other and interact and get to know each other, the stronger the prospects are that we elect leaders who will, one day, pursue peace effectively. I was proud therefore, and grateful, to hear your words, and commend you for them."

Born in 1930 in Vienna, Martin Karplus escaped Austria in 1938 shortly after Hitler entered Vienna. He went with his family to Switzerland, France and finally to the United States, where he attended Harvard (BA) and Cal Tech (PhD). He later returned as a Professor to Harvard, where he developed the molecular dynamics methodology for which he shared the Nobel Prize in Chemistry in 2013.

Supplementary Material

————◦❦◦————

The supplementary material includes:

1. Figures in color
2. Films in Appendix 3

Online access is automatically assigned if you purchase the ebook online via www.worldscientific.com.

If you have purchased the print copy of this book or the ebook via other sales channels, please follow the instructions below to download the files:

1. Register an account/login at https://www.worldscientific.com.
2. Go to: https://www.worldscientific.com/r/q0238-supp.
3. Download the files from:
   https://www.worldscientific.com/worldscibooks/10.1142/q0238#t=suppl.
   For subsequent access, simply log in with the same login details in order to access.

For enquiries, please email: sales@wspc.com.sg.

Index